Designing
Urban Agriculture

Designing Urban Agriculture

A Complete Guide to the Planning, Design, Construction, Maintenance, and Management of Edible Landscapes

April Philips

WILEY

Cover Design: Michael Rutkowski

Cover Photography: Front top: Courtesy of Riverpark Farm, photo by Ari Nuzzo; Front bottom (L to R): Photo courtesy of Atlanta Botanical Garden; Courtesy of Riverpark Farm, photo by Ari Nuzzo; Beth Hagenbuch; Back (L to R) © Beth Hagenbuch; Photo courtesy of Atlanta Botanical Garden; Courtesy of Riverpark Farm, photo by Ari Nuzzo; © Beth Hagenbuch; Courtesy of Riverpark Farm, photo by Ari Nuzzo.

This book is printed on acid-free paper. ∞

Copyright © 2013 April Philips. All rights reserved.

Published by John Wiley & Sons, Inc., Hoboken, New Jersey.

Published simultaneously in Canada.

No part of this publication may be reproduced, stored in a retrieval system, or transmitted in any form or by any means, electronic, mechanical, photocopying, recording, scanning, or otherwise, except as permitted under Section 107 or 108 of the 1976 United States Copyright Act, without either the prior written permission of the Publisher, or authorization through payment of the appropriate per-copy fee to the Copyright Clearance Center, Inc., 222 Rosewood Drive, Danvers, MA 01923, 978-750-8400, fax 978-646-8600, or on the web at www.copyright.com. Requests to the Publisher for permission should be addressed to the Permissions Department, John Wiley & Sons, Inc., 111 River Street, Hoboken, NJ 07030, 201-748-6011, fax 201-748-6008, or online at http://www.wiley.com/go/permissions.

Limit of Liability/Disclaimer of Warranty: While the publisher and author have used their best efforts in preparing this book, they make no representations or warranties with the respect to the accuracy or completeness of the contents of this book and specifically disclaim any implied warranties of merchantability or fitness for a particular purpose. No warranty may be created or extended by sales representatives or written sales materials. The advice and strategies contained herein may not be suitable for your situation. You should consult with a professional where appropriate. Neither the publisher nor the author shall be liable for damages arising herefrom.

For general information on our other products and services, or technical support, please contact our Customer Care Department within the United States at 800-762-2974, outside the United States at 317-572-3993 or fax 317-572-4002.

Wiley publishes in a variety of print and electronic formats and by print-on-demand. Some material included with standard print versions of this book may not be included in e-books or in print-on-demand. If this book refers to media such as a CD or DVD that is not included in the version you purchased, you may download this material at http://booksupport.wiley.com.

For more information about Wiley products, visit our Web site at http://www.wiley.com.

Library of Congress Cataloging-in-Publication Data:

Philips, April.
  Designing urban agriculture : a complete guide to the planning, design, construction, maintenance and management of edible landscapes / April Philips.
    pages cm
  Includes index.
  ISBN 978-1-118-07383-4 (hardback); 978-1-118-33023-4 (ebk.); 978-1-118-33092-0 (ebk.); 978-1-118-33307-5 (ebk.)
 1. Urban agriculture. 2. Edible landscaping. I. Title.
 S494.5.U72P44 2013
 630.9173'2—dc23

2012045251

Printed in the United States of America

SKY10020281_080420

Contents

Preface   vii

Acknowledgments   ix

Chapter **1**

   Food Cities: Ecology + Urban Agriculture   1

     Lafayette Greens, Detroit   1

     Bar Agricole, San Francisco   10

     City Slicker Farms, Oakland   13

     Viet Village, New Orleans   29

     Big City Farms, Baltimore   38

Chapter **2**

   Planning Strategies for Urban Food Systems   41

     Prairie Crossing, Grayslake   41

     River Falls Eco Village, River Falls   51

     Verge Sidewalk Garden, Charlottesville   75

     Scent of Orange, Chongqing   80

Chapter **3**

   Vision, Synthesis, and Form   87

     Villa Augustus, Dordrecht   87

     Miller Creek Edible Garden and Outdoor Kitchen, San Rafael   104

     2001 Market Street, San Francisco   109

     Gary Comer Youth Center, Chicago   117

Chapter **4**

Systems Integration and Connections — 133

Medlock Ames Wine Tasting Room, Healdsburg   133

Our School at Blair Grocery, New Orleans   144

Incredible Edible House, prototype   156

Science Barge, Yonkers, New York   167

Banyan Street Manor, Honolulu   177

Chapter **5**

Lifecycle Operations — 181

Die Plantage, Munich   181

MUSC Urban Fram, Charleston   191

Riverpark Farm, Manhattan   202

VF Outdoors Campus, Alameda   209

Sacred Heart Organic Garden, Atherton   218

Slow Food Nation Victory Garden, San Francisco   224

Chapter **6**

Outreach and Community — 227

Atlanta Botanical Garden, Atlanta   227

Urban Food Jungle, prototype   235

Expo 2015, Milan   239

Alemany Farms, San Francisco   250

P-Patch Gardens, Seattle   253

Glide Church, San Francisco   259

Gotham Greens, Brooklyn   261

Bibliography   267

Image credits   271

Index   273

Preface

DESIGNING URBAN AGRICULTURE is about the intersection of ecology, design and community. It is a dialogue on the ways to invite food back into the city and forge a path towards creating healthier communities and a healthier environment.

When the recession began I planted an edible garden. I started with potatoes and herbs such as parsley, sage, rosemary, mint, and thyme. Within a few weeks I began to notice that even though I only spent 15 minutes a day each morning in the garden my work day stress levels went down and life's hiccups seemed to more easily be put into perspective. My family and I also noticed that our food tasted so much better when I added the home grown edible ingredients to our meals. The potatoes we harvested that first year were the most exquisite and sweetest potatoes I have ever tasted in my life. We all began to eat a bit healthier. If someone was stressing out I'd send them out to the garden to harvest something from our backyard crops and they would come back with a smile on their face and a strawberry or two to share. You could say that we had discovered a little slice of bliss in our daily lives.

What happened next became my adventure into the world of urban agriculture because I wanted to learn more about this incredibly interesting landscape typology and its effects on human health and design. With media headlines such as people around the country being put in jail for planting vegetables in their front yard, or the huge amount of farms failing across the country for various reasons like water and climate, as I dove into the research I began to discover the dark side of our industrial food system and industrialized agriculture. And because the food system in America is broken, the health of our cities and communities are at risk.

As a landscape architect and urban designer I had questions I wanted answers to such as 1. How could these agrarian landscapes be designed at the urban scale to become an integral part of the food system of a city and also be connected to a regional food system, and 2. How could designers collaborate and partner with urban farmers, food entrepreneurs, community organizations, urban ecologists, visionary developers, and city planners in a meaningful way to facilitate the creation of these landscapes while simultaneously addressing issues of human and environmental health, food justice, food security, climate change, cultural aesthetics, and sustainable development. The end result of my quest to answer these questions is this book.

My journey to find answers took many turns along the way. Researching and reading consumed a vast portion of my time tracking down news, articles, books, and web sites that covered the subject matter not readily available in one place. I found seminars, forums and conferences to attend and amazing film documentaries to watch and learn from. One of the most fascinating components of this quest were the conversations I had with colleagues and urban farm enthusiasts from all walks of life about the links between food, design, ecology, and building community.

In particular, my conversations with Jake Voit, who was the Sustainability Manager for Cagwin & Dorward , a top 25 Landscape Contracting firm in the United States located in California turned extremely fruitful. Jake and I had an ongoing conversation thread for over a year's time frame sometimes emailing web sites, articles, ted talks, and sharing video links since we are both passionate about defining the role that designers and citizens can play with urban agriculture in creating positive environments for change. We eventually had enough content on the subject for a provocative dialogue that I invited Jake to present with me at a number of national conferences. By this time Jake was leading the grand vision of the InCommons Initiative for the Archibald Bush Foundation in Minnesota that creates community powered problem solovers. (His work there created highly effective listeners and facilitators of deep relationships based on empathy and a realization of interconnectedness, which influences creating conditions for current and emerging leaders to hold the space for a paradigm shift from individualistic transaction-based communities to shared relationship-based communities.) He began to focus our discussions on the integration of collaborative conversations and ecological parameters into a systems thinking process. I am indebted to the continuing collaborative dialogue we share and the material he has contributed in this book on sustainable agriculture construction practices and how to build resilient communities through collaborative conversations. This includes his description of Cuba's transformation into a sustainable agriculture economy after the trade embargo crisis and his firsthand knowledge of permaculture principles from being raised on an organic farm. His explanations on the intricacies of the soil food web and how to monitor and design for soil health provide clarity and tools for soil management that is a critical component of urban agriculture landscapes. With Jake's background in Environmental Studies, Philosophy, and Permaculture Design, I found his perspective was always unique.

My personal focus then turned towards advocacy and in particular why designers need to play a key role in the integration of urban agriculture landscapes into the urban realm. These explorations and conversations were extended further into physical solutions with my most visionary clients who allowed me to champion urban agriculture within their development projects. I am indebted to them for their trust in letting me design these landscapes for them.

This book showcases projects and designers around the world who are forging new paths to the sustainable city through these urban agriculture landscapes. The case studies demonstrate the environmental, economic and social value of these landscapes and illustrate ways to forge a new paradigm for a greener and healthier lifestyle. The book begins with a foundation on ecological principles and the idea that the food shed is part of a city's urban systems network. It outlines a design process that is based on systems thinking and the design process spheres I developed for a lifecycle or regenerative based approach. It includes strategies, tools and guides to help readers make informed decisions on planning, designing, budgeting, constructing, maintaining, marketing, and increasing the sustainability aspects of this re-invented design typology.

Michael Pollan has said that the garden suggests that there might be a place where we can meet nature half way. Wendell Barry in his *What are People for* essays said that "eating is an agricultural act." My own personal experience with urban agriculture leads me to believe that our dilemma with explaining food as an integral system within the city is because we do not as a culture think of food in this way. People are so disconnected from where their food comes from that thinking about the food system as something they are a part of becomes the first hurdle to tackle if we are going to create positive change. Getting someone to taste food that comes from their own garden is a first step towards optimizing this realization. It is even more rewarding with a classroom of children especially if they have never eaten some of the vegetables or herbs you might get them to taste. These types of local food experiences will begin to change the cultural food beliefs and expand the definition in society to embrace urban agriculture as part of the community's infrastructure systems. If we look at urban agriculture landscapes in this manner we can begin to reduce the amount of people experiencing the effects of a food dessert and increase the ability to foster a more healthy community. The net effect will be to build a more health conscious society that values healthy living as a natural extension of the services a city must provide. My hope is that this book provides a roadmap to anyone interested in the creation and advocacy of edible landscapes that promote beauty, ecological biodiversity and social sustainability in our urban realm.

Acknowledgments

AN ENORMOUS THANK YOU to the grassroot urban pioneers, farmers, chefs, authors, film makers, and design adventurers whose ideas, passion for good food and healthy living, environmental stewardship, and commitment to community and ecoliteracy inspired me to write this book about inviting food back into our cities. Their actions big and small make a difference and are helping to steer urban agriculture into the mainstream as an essential part of urban living.

I am indebted to my friend Jake Voit with whom I was able to share many collaborative conversations with along the way. Thank you for your keen insights on community resiliency, sustainable construction practices and your supportive spirit. This book was seeded from the dinners, chats, and discussions we had about food, design and community. You are a gentle, thoughtful man. Tall too!

I am grateful to my colleagues and fellow urban agriculture enthusiasts around the world who allowed me and my staff to delve into the case studies and resource material featured in this book. I heartily thank you for sharing your thoughts, stories and amazing photos. I especially wish to thank Mary Pat Matheson, Mike Sands, Tres Fromme, Peter Lindsay Shaudt, Margi Hess, Kenneth Weikal, Beth Hagenbuch, Barbra Finnin, Andreas Willausch, Thomas Wolz, Jeff Longhenry, Alexis Woods, Ries van Wendel de Joode, Joe Runco, Marco Esposito, Kevin Conger, Marion Brenner, Wes Michaels, Ted Rouse, Bill Eubanks, Elizabeth Beak, Judith Stilgenbauer, Rainer Schmidt, Gabby Scharlach, Greg Johnson, Eli Zigas, James Streeter Haig, Scott Shigley, Tom Fox, Ari Nuzzo, Hillary Geremia, Pamela Broom, Tony Arrasmith, Peg Sheaffer, Douglas Gayeton, Richard Kay, Rio Clementi Hale Studios, Aidlin Darling Design, William McDonough and Partners, Our School at Blair Grocery, St Croix Habitat for Humanity, Allemany Farms, Big City Farms, Gotham Greens, The Atlanta Botanical Garden, Sandhill Organics, Miller Creek Middle School, and City Slicker Farms. And a fist bump goes to John Liu and Keith Bowers whole lively dinner debate and dialogue at the Bioneers 2011 conference inspired me onward in my journey.

A hearty shout out to my superb assistant Lance Fulton whose methodology oriented mind set was a valuable skill that kept me organized and on time with the deadlines. He tracked, compiled, recompiled, corresponded, and organized all of the case study data files that made it into the book. A big thanks to Ashley Tomerlin who has a keen eye for graphic interpretation of facts and is able to surprise me with her outside of the box thinking. And to the rest of my office for understanding that my late night writing sprees sometimes meant they might not get quick answers or direction from me until I had my organic caffeine.

I am grateful to my editor Margaret Cummins for her patience and support during the final stages of production. Her ability to weed through the vast amount of information I showered her with and distill it into simple succinctness was astonishing.

I appreciate the insights of my initial peer reviewers who helped me guide the development of the original ideas for the book into its evolution. What a process it was!

And to my colleagues who insisted that I include a few of my own sketches in the book or else, I say thanks for the nudging. You know who you are!

To my foodie heroes: Jamie Oliver, Michael Pollan, Alice Waters and First Lady Michelle Obama, you all rock!

Thank you to my mom, Anne, who taught me to love books and gardens. And to my father, Al, who taught me that perseverance is a virtue and hard work is rewarded. I know you are both looking down and smiling on me.

Last, but not least, my family, husband Gary and daughter Gabby, deserve my heartfelt thanks and gratitude for accepting my frenzied nocturnal writing habits fueled by music and dancing during the duration of the writing process like it was the most normal thing in the world, for their understanding when there were weekends I was missing in action, or sitting on the sidelines with my laptop furiously and nerdishly typing away. I can't thank you enough Gary, for being my biggest cheerleader, supporting of all my crazy endeavors and making the meanest, dirtiest martinis with three olives on the left coast. Gabby, girl, you inspire me every day to be a change agent in the world for the next generation and those generations beyond.

CHAPTER 1
Food Cities: Ecology + Urban Agriculture

Lafayette Greens Detroit, Michigan

Though owned by Compuware, a large software corporation headquartered in Detroit, the Lafayette Greens (Figure 1.1) edible urban garden and park looks, feels, and operates more like a public institution landscape (Figure 1.2). Compuware envisioned the project as a means to give back to the community by helping to beautify downtown and creating a space where downtown workers, visitors, and residents can relax and recreate. By making the space an edible landscape, instead of just a plaza, the company is helping to educate the public about health, environmental responsibility, and how to grow food.

Designed by Kenneth Weikal Landscape Architecture, the three-quarter-acre Lafayette Greens is both an aesthetic and functional success, winning an honor award from the American Society of Landscape Architects in 2012.

Figure 1.1 The vacant city lot was transformed into an urban agriculture oasis.

Figure 1.2 The urban agriculture plaza situated in downtown Detroit.

1

The site design incorporates a wide variety of elements and materials to maximize programming on the site and foster strong connections (Figure 1.3) outwardly to the landscape's urban downtown context (Figure 1.4). In addition to the custom metal raised beds (Figure 1.5), elements include garden sheds (Figure 1.6), a children's garden and learning area, and a dedicated space for public art. Informative signage serves to educate the public about the connections between horticulture and sustainability.

Figure 1.3 Design and materials were employed to create a functional and aesthetic space that connects into the urban fabric through circulation and programmable elements such as these repurposed steel beams.

Figure 1.4 The materiality references the surrounding urban context.

Figure 1.5 Local volunteers work the raised planting beds.

Figure 1.6 Fun storage sheds reference the vernacular farm landscape.

Food Cities: Ecology + Urban Agriculture   3

Built on the lot of a recently demolished building, the site's geometry is based partly on the desire lines of those who traversed the site while it was vacant. This facilitates and encourages circulation through the site rather than around it. A wide main walkway lined with lavender (Figure 1.7) and custom-built benches traverses the site. Lavender was chosen because it has been shown to have a calming effect on people. The site layout and edible-plant-bed orientation were designed to maximize sun exposure for the site-specific shade patterns caused by surrounding buildings. The children's area is in one of the sunnier spots of the plaza to promote lingering (Figure 1.8). And unlike many projects, the garden's aesthetic geometry can be appreciated by those who look down on it from nearby buildings.

The garden is intricately detailed, and incorporates many reused and repurposed materials. Concrete rubble is used to form gabion curbs, while broken sidewalk pieces serve as pavers.

Figure 1.7   The main artery path is lined with lavender for its calming effects.

Figure 1.8   Repurposed steel drums are used in the children's area.

On-site garden sheds are built from reused wood and salvaged doors. Repurposed food-grade steel drums are used to make smaller planters in the garden's children's area (Figure 1.8). Environmental efforts extend beyond materials, with the site's stormwater being captured, filtered, and detained in a bioswale of native plants. Adjustable drip irrigation is used to tailor water consumption to each plant's needs, minimizing waste. Roughly 70 percent of the site's surfaces are permeable, including small, drought-tolerant fescue lawns.

The all-organic urban garden is highly productive as well. Over 200 species of plants are grown on site, including vegetables, berries, herbs, and even kiwi vines on an overhead trellis. An "orchard meadow" of native fruit trees and an heirloom apple orchard line the site's northern edge. Currently, the garden is run by a garden manager from Compuserve, and worked by volunteers, many of whom are Compuserve employees. All food is donated to downtown Detroit's local food banks, with the volunteers allowed to take home a little food themselves in gratitude for pitching in.

Design Team:

Client/Developer:	Compucare
Landscape Architect:	Kenneth Weikal Landscape Architecture
General Contractor:	Tooles Contracting Group, LLC
Landscape Contractor:	WH Canon Company
Irrigation Consultant:	Liquid Assets, LLC
Architectural Consultant:	Fusco Shaffer and Pappas Inc.
Garden Shed Construction:	Mackinac Woodworking Concepts
Civil Engineering:	Zeimet Wozniak & Associates
Structural Engineering:	Desai/Nasr Consulting Engineers
Electrical Engineering:	TAC Associates, LLC

Designer as Change Agent

The twenty-first-century sustainable city requires the merging of urbanism with sustainable food systems. The design strategies for agricultural urbanism are about reinviting food back into the city and reconnecting people with their local and regional food systems to promote a healthier *and* more sustainable lifestyle. This challenges today's industrial food system that currently separates people from their food sources. Urban agriculture is, now more than ever before, a movement in transition, and these new urban landscapes are demonstrating that they are far more than growing vegetables on abandoned lots.

In addition to needing water, food is a basic human need for human existence. Food is also essential to economic growth. Food provides a new perspective for answering the question about how we make our cities more livable places. Everyone needs food, in all probability likes food, and has shared food with others. Food represents both celebration and sustenance in all cultures. Almost everyone has a personal history that centers on rituals and relationships that revolve around food. Whether birthdays, family holidays, or even meeting for dinner with friends, food plays a large part in our daily lives. Food, because of its universal appeal and appreciation, can become an important key to further advance the sustainable city dialogue.

Along with integrating a more comprehensive ecosystem-based approach to the redesign of our cities and towns to handle the ever-increasing complexity of the urban realm, integration of an economically viable urban food system needs to become an integral part of the urban ecosystems that frame the foundation of the sustainable city. The time has come for designers to act as change agents and design for integration of natural systems with urban systems into city infrastructure. That infrastructure needs to include urban agriculture as an integral part of an economically viable food system within a city.

An economically viable urban food system would result from an *ecological and biological* based city-planning model that would focus on health (human and city), community (support and connectivity), and ecosystems (natural and manmade). Current urban design and planning is focused on the fragments rather than a cohesive whole. A new way toward designing integration is emerging through ecological-based urban agriculture. This integrative process, also known as integrated systems thinking, focuses on solutions based on the interconnectedness of the systems as a whole unit, rather than separate units.

This book will outline a framework of information to aid in the creation of urban agriculture landscapes that promote ecological biodiversity and social sustainability. Consideration for creating these landscapes needs to accommodate design strategies that integrate social, ecological, and economic values to achieve the best results. Plus, diving deeper into planning and policy, information will unfold on how to incentivize and design a regenerative landscape that benefits the community and local ecology. An emphasis now must be on how designers are the change agents for this new green infrastructure in today's cities.

In an interview in the *Yale Daily News* in October 2011, former Slow Food USA president Josh Viertel said, "Slow Food is like a gateway drug for civic engagement, environmentalism, for changing the world, because when you share food with people you see what you have in common. Food becomes a vehicle for understanding each other and for dealing with issues of race, class, oppression, and gender." Thus, food and the conversation around food becomes the topic starter for the community and civic dialogues that must ensue.

In recent years, there has been a tremendous upsurge of interest in growing food in urban centers. In the past quarter century, food production has been pushed to the periphery of the city. Now the trend is to connect it back to the heart of the city and build bridges between the urban communities and peri-urban and rural communities. This resurgence is in response to concerns about rising food prices, food miles, and the environment. It is also because people want better access to good, healthy, and affordable food, and to enjoy cultivating beautiful green spaces and meeting local people. Eating food you have grown yourself is a visceral experience. There are now millions of people planting urban farms today in the United States. The trend is up 35 percent over the past year and 15 percent of the United States now has a backyard garden. In a country of 300 million people, this is a huge trend. You would think that this means that the new urban farmers are hip twenty-something-year-olds, but in actuality, 80 percent of US farmers today are over 60 years old. What will happen in 10 to 15 years when they need to retire? There are an estimated 14 percent of people in the United States who experience food insecurity every day. This means that there are approximately 49 million people in the United States alone who do not know where their next meal is coming from.

The slow food movement with its emphasis on natural, healthy, nonprocessed foods has spawned a whole new emphasis on connecting people on a personal level with what they are eating. The promotion of the "seed to table" and "farm to table" local food concepts has started a new locavore movement that promotes the locavore lifestyle as the preferred sustainable choice. This includes the most recent trends for urban agricultural landscapes cropping up in cities: high-density mixed-use projects, community supported agriculture, otherwise known as CSAs, urban rooftop farms, the return of civic victory gardens, edible green schoolyards, restaurant gardens, corporate community gardens, and others.

On the zoning front, the major challenge in some cities is in obtaining the legal right to grow vegetables in your own yard. This is not possible in many cities and towns across America due to local zoning laws and planning policies of another era that forbid growing food in urban areas. Other cities such as Detroit, Chicago, Baltimore, New York, Seattle, and San Francisco are setting new precedents in reforming these outdated codes to welcome urban agriculture into their cities on a larger scale. Some city planners are even considering the potential of new agricultural districts including food production within the city limits, but with every gain, there are also pushbacks and setbacks.

Today's urban farms have not yet found a way to thrive in the market economy when a majority rely on volunteer labor and grant funding. While at the epicenter of ecological sustainability, they have yet to reach a level of economic sustainability that would provide impetus for the trend to reach a tipping point level for wider traction and viability. Along with integrating a more comprehensive ecosystem-based approach to the redesign of our cities and towns to handle the ever increasing complexities of the urban realm, integration of an economically viable urban food system will need to become an integral part of the urban ecosystems that frame the foundation of sustainable cities. An economically viable urban food system would result in an *ecological- and biological*-based city planning model that would focus on the health of the city and the human beings that reside in it, provide a framework of support and connectivity for communities to flourish and prosper, and integrate an infrastructure that manages resources through an ecosystem services approach that builds resilience and regenerative environments.

The time has come for designers to act as change agents and design for the integration of natural systems with urban systems into city infrastructure of which urban agriculture would be part of a food system network. Current urban design and planning is focused on fragments and pieces rather than a more cohesive whole. A new path toward designing integration is slowly emerging in some planning circles through the lens of ecological-based urban agriculture. This integrative process, also known as integrated systems thinking, focuses on solutions that are based on the interconnectedness of the systems as a whole unit rather than the separate units.

The project case studies in this book demonstrate that designers are taking action and are making changes through food landscapes that reflect an integrative systems-oriented approach that may help provide a guide for building a more sustainable city. This book will illustrate how current projects and urban ag landscapes around the globe are transforming our ideas on the integration of food into the city.

Nature + City

The romantic ideas of living in harmony with the city inspire artists, writers, and visionaries to create images, narratives, and models that depict what this might mean for the twenty-first century and beyond. This conceptualization includes addressing our current increasing disconnect from what we eat, the continued overdevelopment of our cities at the expense of nature and human health, and the increased destruction of local, regional, and global natural resources. The exploration of what it means to live in harmony with nature in the city is fuel for innovation, technologies, and ideas that may have the genesis for contributing to a new paradigm for designing the sustainable city. People are becoming more in need of ritual and relationships. Charles Montgomery in *Happy Cities* builds the case that people need each other and they also need contact with nature to be happy. But how much nature do we need in a city to be happy? What does this nature look like and how does it become a more integral part of city pattern and identity?

Romanticism of Agriculture + City

When you create an edible garden or urban farm or another type of urban agricultural landscape, you learn right away that it is not as romantic as it sounds. It is actually a lot of hard work! The roots of the world's agricultural society and the invention of farming were based on the growth of wheat, rice, and maize crops, though hunter-gatherers were around at least 35,000 years before the expanding farming culture changed civilization (Standage 2010). Historical indications of urban agriculture go back only as far as 10,000 years ago, where civilizations in Egypt, China, and India illustrated farming as part of the daily lifestyle in the remnants that have been preserved and recorded by anthropologists and scientists. From these early times in various societies around the world, urban agriculture was manifested as a contained and controlled zone (Figure 1.9) within the compounds and cities by thick, sturdy walls to keep out the untamed wilderness located just outside the city walls. Nature was wild and found outside of the city, not within the city.

Figure 1.9 Agricultural practices have evolved overtime and paralleled society's changing relationship with nature.

Garden Cities

Fast forward to the nineteenth century when a movement was begun for *garden cities,* a method of urban planning initiated in 1898 by Sir Ebenezer Howard in the United Kingdom. This proposal favored a decentralized city layout sprinkled with an abundance of public parks and pastoral open space that included orchards and was laid out in a radial pattern (Figure 1.10) with wide boulevards and spatially differentiated land uses. It was a pastoral view of the city and the opposite of overcrowded and dirty cities of the time. The city is shown as a centralized site of 1,000 acres and surrounded by agricultural land of 5,000 acres to support a city population of 32,000. Within the centralized city, there would be parks small and large, orchards, small dairy farms, and other types of productive landscapes. The proposal was written as a business model; some contemporary concepts—such as transit-oriented development and urban growth boundaries—can be attributed to Howard. While not well received due to its perception as a utopian socialist ideal, Howard's overall goal for the garden city was to bring nature back into the city. In this respect, Howard succeeded, because integrating productive landscapes back into the city is what is currently going on with the urban agriculture movement in cities today—and not just in the United Kingdom.

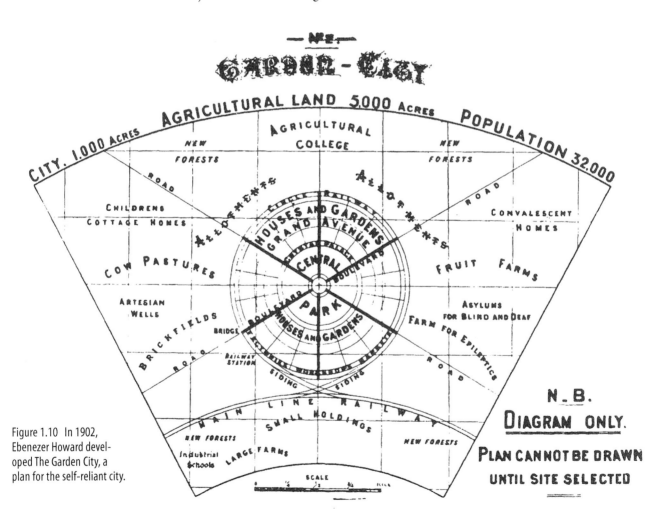

Figure 1.10 In 1902, Ebenezer Howard developed The Garden City, a plan for the self-reliant city.

In the United States, garden city principles greatly influenced developments such as Pittsburgh's Chatham Village; Reston, Virginia; Garden City, New York; and other new towns planned in the early part of the twentieth century. Greenbelt towns such as Greendale, Wisconsin; Greenbelt, Maryland; and Greenhills, Ohio, were planned between 1935 to 1937, as self-contained communities surrounded by greenbelt parks containing proportionate area of residences, affordable housing, places to work, open space, and agricultural land. These developments served as experiments in innovative urban planning. The underlying garden city principle was to produce economically independent cities with short commute times along with the preservation of natural, open-space countryside that was both pastoral and productive. Garden city principles also influenced the design of colonial and post-colonial capitals such as New Delhi, India, and Canberra, Australia, during this period. Unfortunately, garden suburbs were also created at the same time, which had the opposite effect because they were located on the outskirts of cities without allowing for industry; thus forcing residents to rely on transportation to commute into the city for work.

The garden city principles offer some insight into developing more sustainable communities and cities that incorporate urban agriculture into its open space infrastructure. Related urban design concepts influenced by this movement include transition towns and transit-oriented developments.

Growth of the Farm to Table Movement

So let's fast forward to the present. Urban agriculture is a movement in transition. About 75 million people tend virtual gardens in the game Farmville on the Internet! We are also the most obese society this planet has ever seen. We obviously have an affinity for food. The resurgence in urban agriculture focused on local food access is based on the recognition that the industrial agricultural system does not add to a human being's quality of life. Slowing down to plant your own edible landscape, or taking part in a community garden, or buying from a local farmer at the market is connecting us back to our historic roots of local and meaningful food production.

One of these local food movements that have received a lot of publicity—including high-profile members such as chef and cuisine innovator Alice Waters—is called *slow food*. Slow food is an idea, and it is all about the promotion of a more healthy way of living and eating. It is both a global and a grassroots movement composed of thousands of members around the world with a primary focus on linking the pleasure everyone experiences through food with a stronger commitment to the betterment of the community and the environment.

Another high-profile champion of healthy, local food comes from First Lady Michelle Obama's mission to tackle childhood obesity. The reinstallation of the White House organic garden on the front lawn of the White House and her Let's Move campaign, which focuses on educating elementary students on the need for physical activity in their daily lives, are both shedding light on the strong links between organic, locally grown produce, physical activity, and the health of our children.

Bar Agricole San Francisco, California

The name of Bar Agricole, in San Francisco, California, references both the *agricole* variety of rum, featured in one of its signature cocktails, and the notion of agriculture itself. The restaurant and bar go beyond paying lip service to these notions, putting edible plants at the center of both its image and its culinary exploits.

The restaurant is located in a historic building that once housed the Jackson Brewing Company in the more industrial South of Market District of San Francisco known as SOMA to locals. First constructed in 1906, the unfinished building was destroyed by the earthquake of the same year, and reconstructed in 1912. The building was renovated by Aidlin Darling Design in 2011, becoming San Francisco's first LEED Gold building. Part of the sustainable vision was to have a restaurant on the ground level that was rooted in sustainable farm-to-table philosophy and compatible with the sustainability values of the owner. Sustainable strategies (Figure 1.11) included regional and on-site fabrication within a 15-mile radius; the restaurant to benefit from the building's solar array and living roof; natural ventilation for passive cooling with cross-ventilation; permeable pavers for outdoor dining and rear alley parking spaces; recycled and reclaimed wood materials used for chairs, benches, and beams; plus the on-site agriculture component that provides produce for the restaurant's use.

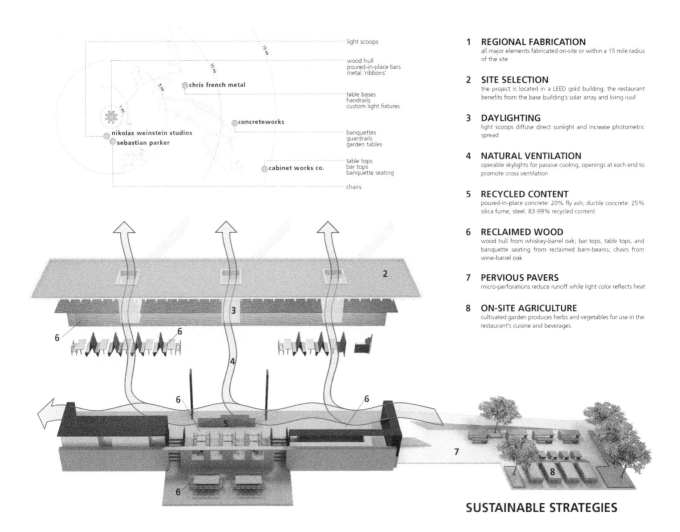

Figure 1.11 The restaurant's passive design technologies have eight key components.

Food Cities: Ecology + Urban Agriculture 11

Figure 1.12 Wood planters are used for herbs for culinary accent in the kitchen and bar.

Figure 1.13 The materials reference the industrial nature of the surrounding neighborhood.

Figure 1.14 The canvas opening rolls up on a sunny day.

Though the restaurant's interior is beautifully detailed, it is the outdoor dining space that speaks to the agricultural name and makes the sustainable-minded restaurant more unique. The space has an intimate garden feel, thanks to 500 square feet of wooden raised planting beds (Figure 1.12) that serve as the highly fragrant and living backdrop for diners. The raised beds contain a variety of herbs such as mint, lavender, and rosemary, and citrus trees such as lemons and limes, which are used in Bar Agricole's food and artisanal cocktails. Other leafy greens and vegetables are seasonally introduced to take advantage of seasonal influences to enhance the culinary and visual experience.

The recycled and locally sourced materials (Figure 1.13) provide a rugged anchor tying the project setting to its industrial warehouse roots. The reclaimed cedar wood walls of the garden court provide warmth and light glows at night from a series of built-in light panels within the walls. A canvas overhead awning (Figure 1.14) is pulled open on warm, sunny San Francisco days and beautiful starry nights, yet it can provide light cover from the typical foggy nights and rain. The walls provide a tranquil setting from the busy street just outside the walls, and the garden is further removed from the noise of the street by the sound of the vegetated fountain.

Design Team:	
Restaurant Owners:	Andreas Willausch, Thaddeus Vogler, and Eric Johnson
Chef:	Brandon Jew
Architect:	Aidlin Darling Design
Contractor:	Northern Sun Associates
Sustainability Consultant:	Simon and Associates
Garden Fabricator:	Cronin Construction & Development, Inc.

Addressing Food Justice and Food Security

In its simplest terms, food security means access to safe and fresh food at all times by every person. Urban agriculture allows for the creation of increased local sources to meet this need.

> In the United States, 49 million people, out of the 300 million total population, are unsure where their next meal is coming from. That is about 15 percent of the population, a statistic that is not going unnoticed and one that many urban farms are trying to address.

They are tackling tough issues such as access to healthy food, childhood obesity, and limited access to food for low-income, underserved communities where grocery stores might be as far as five to ten miles away and transportation is limited. These farms are operating under the premise that healthy and affordable food is a basic human right.

There has been a resurgence of agriculture in blighted cities such as Detroit, Baltimore, and New Orleans. Detroit's guerrilla gardeners began by reclaiming patches of vacant land in the name of fresh food. These landscapes are concerned with food justice and offering people who live in food deserts, where there is no access to healthy fresh food, a chance to incorporate it into their daily life. The problem areas offer an opportunity for new food infrastructure and business models to emerge. It is an opportunity for a new food economy.

As industrialized agriculture became the dominant method of food production in the United States, the economic functions of production, distribution, and retail sales have focused on the bottom line of profits, not accessibility. Grocery stores and food markets are prominent in affluent communities with plenty of money to spend on a variety of healthy foods. The options for consumption increase as the amount of money in the neighborhood improves. Whole Foods, a store that promotes healthy food to its mostly affluent customers, is rarely found in poor communities. The economics of food does not incentivize retailers to provide fresh, healthy unsubsidized foods in poor neighborhoods, where the amount of money they can spend on food is less.

Numerous trends have contributed to these food deserts—places where good-quality food is scarce—including the concentration of wealth, the shrinking middle class, and fewer neighborhoods that can support a variety of food stores. Big-box grocers such as Super Target and Walmart are typically far away from poor urban communities. The lack of subsidies for fresh healthy foods cannot compete with subsidized highly processed foods based on corn and soybean production, GMO seeds, and nonsustainable methods. What results are urban communities with few if any grocery stores able to stay in business selling expensive fresh foods shipped, on average, 2,000 miles or more.

One example of a response to a food desert is taking place in Oakland, California, where these food deserts are a huge challenge and huge opportunity to regenerate the local food, community, health, ecological, and economic systems. Here, people are using the land that has been unused for agriculture for years, reclaiming their urban environment, and taking their food and health into their own hands. Members of the community are partnering with their neighbors, nonprofits, city officials, and entrepreneurs and working as change agents to move through the many challenges of using property for food production that was not designated for that purpose in city planning.

City Slicker Farms, Oakland, California

West Oakland has many challenges to overcome for a vibrant urban food system: toxic soil, little public land for farming, industrial landscape, lack of biodiversity and habitat, no markets for selling, low rates of education, and very few home owners who can use their land as they wish. Many of the ecosystem services of a healthy urban landscape are gone, with little vegetation or common spaces with good soil, water runoff management, clean air, beneficial insects, habitat for birds and native plants, space for perennial food crops like trees, and so on.

In 2001, a group of community members got together to address this concern by growing their own food in vacant lots, the first one being Willow Rosenthal's donated lot. Interested members of the neighborhood volunteered to grow the food and then sell it at Center Street Farm Stand and share the rest for donations. The effort has grown to include over 100 backyard gardens, seven Community Market Farms, weekly farm stands, a greenhouse, and urban farming education programs. The organization is now known as City Slicker Farms. Today, City Slicker Farms has grown into a large organization with a variety of ambitious goals, projects, and initiatives. Its efforts are organized into three main programs: the Community Market Farms Program, the Backyard Garden Program, and the Urban Farming Education Program (Figure 1.15).

Figure 1.15 City Slicker Farms aims to educate residents on growing their own food and nutrition.

Figure 1.16 Community residents growing their own food.

The Community Market Farms Program continues the legacy and mission of City Slicker's very first farm—producing high volumes of food on underutilized land for low-income community members. The produce is distributed at a weekly farm stand on a donation-only basis, allowing residents to pay what they can afford and assuring that none are denied access to healthy food. The program generates its own compost for the gardens, mixing donated sawdust and manure with food scraps from local businesses and residents.

The Backyard Garden Program takes the food justice initiative of City Slicker Farms one step further, empowering low-income Oakland residents to be more self-sufficient by growing their own food in their own backyards. The program is free and provides both resources and training to the budding urban farmers. The process begins with a soil test of the yard to be cultivated, to see what can grow well there. City Slicker staff members then meet with backyard gardeners to discuss what crops the gardeners would like to grow, and what is possible in their location. A garden plan is then drafted, and the garden constructed by the backyard (Figure 1.16) gardeners, City Slicker staff, and volunteers. Mentoring is provided for the first two years, along with soil, plants, and other supplies. Gardeners are encouraged to share what they have learned with family and friends, and can become mentors themselves after one year in the program.

Part of City Slicker's mission is to train the next generation of urban farmers and urban agriculture leaders. It does this through a number of components, which are grouped together in its Urban Farming Education Program. One component of the program is its allyships, which are similar to internships, but available to people of all levels of experience. Participants help with both the Community Market Farms Program and the Backyard Garden Program, while learning about horticulture, construction, garden management, marketing, and fundraising. For school-aged youth, these same skills can be learned in the Youth Crew summer program (www.cityslickerfarms.org/youthcrew), which provides training and a monthly stipend.

To help support these programs, City Slicker operates a greenhouse at a local high school. Over 30,000 seedlings are grown per year for use in the Market Farm and Backyard Garden Programs, and for sale to the public. Additionally, the greenhouse is used to teach horticulture, environmental science, and nutrition for City Slicker's Urban Farming Education Program, and for the school's own science classes.

In 2010, City Slicker Farms was awarded $4 million to build the West Oakland Park and Urban Farm, which is to be the flagship farm of the Community Market Farms. It will be the largest farm in the program, and the only one owned by City Slicker. The funding comes from the state of California's Proposition 84 bond initiative, which awards

Food Cities: Ecology + Urban Agriculture   15

funds for projects involving natural resource protection and parkland improvements. The park design came from a three-month community design process, through which input was solicited from community youth, families, seniors, and other stakeholders (Figure 1.17). The input was integrated into a unified plan (Figure 1.18) by CMG Landscape Architecture. It will function as a park as well as a farm, so site elements will include a lawn area for play, a vegetable patch, a fruit orchard, a chicken coop, a beehive, a dog park, and a playground. There is also space dedicated to a community garden in which families and individuals can lease a plot to grow their own produce.

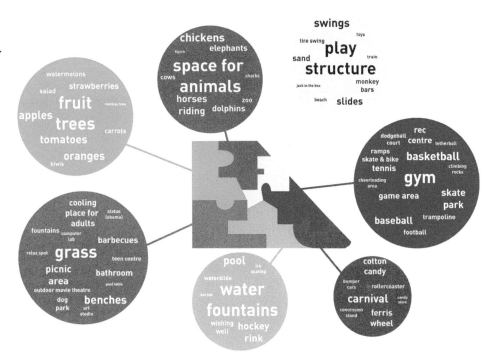

Figure 1.17 Community defined programs for the new urban farm and park.

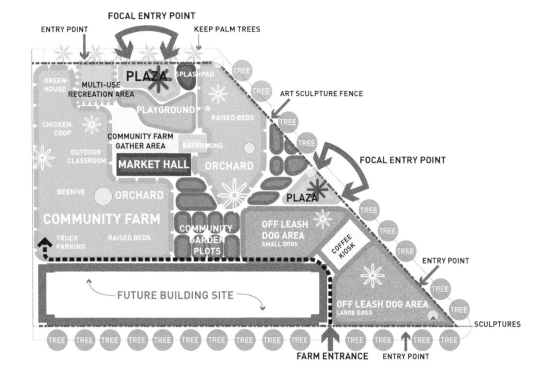

Figure 1.18 Site elements for the proposed urban farm and park.

West Oakland Park and Urban Farm Design Team:

City Slicker Farm:	Barbara Finnin, executive director
Landscape Architect:	CMG Landscape Architecture
Project Management:	mack5

A sampling of urban farms focused on food justice:

The Food Project, Lincoln, Massachusetts

Added Value, Redhook neighborhood, New York City

Urban Farm, Phoenix, Arizona

City Slicker Farms, Oakland, California

People's Grocery, Detroit, Michigan

Growing Power, Detroit, Michigan

Our School at Blair Grocery, New Orleans, Louisiana

Hollygrove Market & Farm, New Orleans, Louisiana

Dig Deep Farms, Castro Valley, California

Urban Adamah, Berkeley, California

Veggielution, San Jose, California

Tenderloin People's Garden, San Francisco, California

Farm + Food Lab, Irvine, California

Human health, childhood obesity, ecological medicine

In a recent study, all newborns in the study were born with organophosphates in their first bowel movement, and most had six or more pesticides. How do we create value for organic landscapes to eliminate this type of toxicity in human beings? Food can be a good reason for clients to want to make a change over from traditional chemical-based maintenance to organic maintenance.

Living Downstream is both a book and film about the story of Sandra Steingraber, an ecologist and cancer survivor. The film by Chanda Chevanne is a thoughtful documentary that brings awareness to the dangers of pollutants in our lives and based on Sandra's book of the same title. In her book, Steingraber outlines the chemicals we deal with on a daily basis, from birth, that have consequences on our health. These are chemicals that are used for all sorts of products and materials all around us. Just in the umbilical cord alone, medical tests have found 287 different chemicals! Steinberger names the sources of a few of these chemicals that include pesticides, stain removers, wood preservatives, mercury, and flame-retardants. We are exposed to thousands of chemicals in our environment, and most of us are unaware of them. Per year, 22 chemicals in our environment are neurotoxins and another 1,000 are suspected of affecting the nervous system. These chemicals can affect the development of a fetus during development and cause premature births. Benzo[a]pyrene, a chemical ingredient in tobacco smoke, diesel exhaust, and soot, has been found to damage eggs in the ovaries. Exposure to pesticides can reduce sperm count in men.

Several experts in the fields of toxicology and cancer research make important cameo appearances in the film, highlighting their own findings on two pervasive chemicals: atrazine, one of the most widely used herbicides in the world, and the industrial compounds known as polychlorinated biphenyls (PCBs). Their scientific work further illuminates the significant connection between a healthy environment and human health. The film itself follows Steinberger's journey and the chemicals journey that she is fighting. Chevanne says about the film, "We follow these invisible toxins as they migrate to some of the most beautiful places in North America. We see how these chemicals enter our bodies, and how, once inside, scientists believe they may be working to cause cancer." Through thought-provoking visuals and a unique storytelling style, the film presents the links between toxins in the environment and the negative impacts they have on human health in a manner that film critics have called powerful and haunting.

Equating calories with higher food prices is another way to illuminate the obesity problem in America. According to the Department of Agriculture's 2009 report, an average American consumes 600 calories a day more than people did 40 years ago (United 2010). Additionally, childhood obesity has more than tripled in the past 30 years. The prevalence of obesity among children aged 6 to 11 years increased from 6.5 percent in 1980 to 19.6 percent in 2008. The prevalence of obesity among adolescents aged 12 to 19 years increased from 5.0 percent to 18.1 percent. These findings are identified by the CDC in its 2009 report on Childhood Obesity. (Centers 2009)

These 600 calories in our diet are generally taken into our bodies from high-fructose syrup and other additives that previous generations have not been subjected to. These chemical compounds continue to be raised annually and replace the nutritional value in processed foods, adding more invisible nutritionless fillers to the average diet. "The postwar farm policy has been to produce as many calories as possible," says David Wallinga, who is the director of the food and health program at the Institute for Agriculture and Trade Policy in Minneapolis. The result of producing as many calories as possible is that we are seeing more corn, soybeans, and wheat made into processed sweeteners, oil, and flour bulking up supermarket foods, and supersized meals as mainstays in the American diet.

How many people understand that chemicals like PCBs raise the risk of premature birth? How many people understand the links between obesity and processed foods? How can we make decisions as policy makers, consumers, planners, and citizens when our ecological understanding is limited? One way will be to increase our ecoliteracy as a nation.

KEY CHEMICALS TO AVOID

Bisphenol A (BPA): BPAs are chemical compounds that are added to plastics to make them more durable. They are commonly found in the linings of food containers, beverage cans, some baby bottles, and drinking bottles, and they can leach out into foods and liquids. Harmful side effects can include increased risk of breast and prostate cancers, infertility, polycystic ovary syndrome (PCOS), and insulin resistance, thought to lead to type 2 diabetes. To protect yourself from BPA, it is recommended that you use glass containers for food, drink from stainless steel water bottles, and avoid canned foods or look for cans and bottles that say they are BPA free.

Polychlorinated biphenyls (PCBs): These chlorinated chemicals that are typically used as coolants and lubricants in electrical equipment and old fluorescent light fixtures and appliances are major endocrine disruptors. In 1977, because of evidence showing that they have built up in the environment and caused harmful health effects, their manufacture was banned in the United States. However, because of their persistence in the environment, farmed salmon and a variety of freshwater fish have been found with PCBs still circulating in their systems. Check out safe fish lists created by institutions such as the Monterey Bay Aquarium to make sure that the fish you buy and eat are PCB free.

Volatile organic compounds (VOCs): These are chemical compounds that are emitted as gases by seemingly innocent everyday products as paints, plastics, cleansers, air fresheners, dry-cleaning, and cosmetics, and they can cause problems to people's endocrine systems. Types of effects that can be experienced are nausea, headaches, drowsiness, sore throat, dizziness, and memory impairment, to name a few. It is also thought there is a possible link to cancer. For kitchen-cleaning projects, the recommendation is to switch to all-natural products, such as baking soda, hydrogen peroxide, white vinegar, and lemon juice. There are also a variety of eco-friendly paints produced without VOCs in your local hardware and home-improvement stores.

Chlorinated products: These products include everyday household items such as white paper towels, napkins, and coffee filters. The white color is the effect from bleaching with chlorine. Most people do not know that these products affect human endocrine systems and hormones in a negative way. The Environmental Protection Agency has found that dioxins, the byproducts of many industrial processes involving chlorine, are 300,000 times more carcinogenic than DDT, a synthetic pesticide now banned in the United States since 1972. To best protect yourself, select products labeled chlorine-free, or PCF.

Cosmetics: See also "Dirty Thirty," a list of ingredients to avoid in daily use products for beauty and personal care created by Teens Turning Green, a nonprofit organization built on targeted, collaborative, student-led campaigns for change (see http://wordpress.teensturninggreen.org/wp-content/uploads/2012/02/TTG-DirtyThirty.pdf). The organization is guided by Executive Director Judy Shils and located in Marin County, California. One of the campaigns is focused on sustainable food awareness through programs such as Project Lunch that aims to rethink and transform school lunch programs in Marin County.

Peak Resources and Their Effect on Urban Farming

The dependence of our food system on fossil fuels will make peak oil a turning point in how our food is grown, distributed, and valued. Petroleum is a nonrenewable natural resource because of its finite limitations available on the scale that can sustain its current global consumption rate. The concern with the depletion of the earth's finite reserves of oil and its effects on a society dependent on it is what is known as the field of peak oil. Peak oil reflects the point at which the maximum rate of fossil fuel extraction has been reached after which the rate of production will begin a terminal level of decline. There is an active ongoing debate on how to measure peak oil but optimistic estimates forecast that a global decline will begin after 2020 assuming that major investments in alternative energy sources begin to occur now. Currently the United States spends 15 percent of energy use on feeding Americans. It is the types of food we eat, how we farm it, and how we buy it that has the greatest impact on the amount of energy used. This overdependence on processed foods, red meat, and the reliance on pesticides impacts not only higher energy use, but also negatively impacts human health. It will require everyone to take part in the evolution of the present food system to move it from large, industrial farms to small, sustainable-oriented urban food systems. Many people in the urban setting will need to vote for city rules to allow urban farming, source food from farmers in the local area, develop robust local food economies, and integrate new uses for land into city planning. Many land options and new skill sets will be needed due to the decentralization of the food system to urban lands. We will need many small farms and farmers growing sustainably and who can get food to market without relying on high uses of fossil fuel systems. Peak oil will increase the price of production and require our food to be grown with less fossil fuel–based chemicals and closer to the site of sale to reduce the need for fuel. Using large farm equipment and semi-trucks for transporting food an average of 1,200 miles will no longer be economically feasible.

Figure 1.19 Cafes and markets are integrated into public spaces throughout Havana.

Peak Oil and Cuba

As the Soviet Union collapsed in 1991, it lost its close relationship with Cuba (Figure 1.19), resulting in the discontinuation of subsidized oil supplies to Cuba. Cuba's industrialized agriculture, health care, education, and transportation systems could not be sustained as they had been. Each of these sectors was highly centralized into large systems that relied on fossil fuels–based transportation, distribution, mobility, and power. The systems worked well as long as cheap oil was available for the business model to sustain itself. Centralization makes sense when oil is abundant and cheap. Similar to the United States, a suburban model of development, a centralized system must rely on cars and vehicles for shipping products across the country to make it work.

When Cuba lost its cheap oil supplies, it entered what is now called the Special Period of decentralization of its infrastructure into a diverse and local infrastructure (Figures 1.20 and 1.21). Through the Special Period, Cuba changed from centralized clinics to having doctors and clinics in most communities throughout Cuba, and shifted from 3 universities to about 50. Semi-trailers became community buses, and industrial farms became organic local urban ones. The shipping of products and transportation by car become impossible for most Cubans. Most things had to be produced and provided within walking distance of all the communities across the country. People, on average, lost 20 pounds, lost most of the meat in their diets, and had rationed foods, as the industrial economy based on cheap oil collapsed.

Figure 1.20 Cuba has developed a strong urban agriculture economy within its cities.

Figure 1.21 Small-scale farms are opportunistically placed on vacant lands.

Figure 1.22 Local markets thrive with local produce and products.

In order to cope with this situation, the government made important policy choices to ensure the country could create a food system that could feed its people without the use of fossil fuels–based technologies. The government also asked for help from the International Permaculture Design community to help in designing the sustainable agriculture, education, and mentoring programs that were now necessary because there were not enough farmers to execute a decentralized system of many small farms throughout the entire country. Large, tractor-based farms were no longer an option. Each person who was interested in using land for farming was given some to farm (Figure 1.22); if that person did not farm it, it was given to someone else. This was done everywhere, from small towns and throughout Havana. Synthetic pesticides were also no longer available, so the farming became organic. Waste from communities did not get shipped but was composted in large urban worm bins to recycle the resource for fertilizer. Farmers' markets were set up in throughout the cities to ensure that people had a place to exchange goods within walking distance of the homes, farms, and a central location.

According to recent statistics, 60 percent of the food eaten in Havana is grown in the city limits (Figure 1.23). Most of the agriculture is organically grown. Many people are now able to farm land when before they did not have access to those lands. The most important career in Cuba became farming, due to the shift from a fossil fuel industrial economy to an agrarian economy. Cuba's Special Period is a great model to learn from when thinking of the United States' highly centralized system of food

Figure 1.23 In Havana, 60 percent of the food grown is grown within the city limits.

production and distribution. The current system is effective in creating a lot of food and jobs, but it is not resilient to a decline in oil availability and price increases that are beginning to affect the global economy. Ideally, communities would start shifting now to a more decentralized food system before they are forced to transition quickly, like Cuba did. As the price of fuel rises, our food system could be made more resilient by providing food in a variety of ways. Large farms are good for some crops, and urban farms are good for others. Also, if small farms were to receive similar subsidies of larger operations, then the viability of small-scale businesses could be similar to large ones. The lessons from Cuba's sudden shift from oil dependence can be applied to any community whose infrastructures are based on the abundance of cheap oil. Cities will need to plan and execute this transformation as an integrated community with all stakeholders and sectors working together to decentralize the food system infrastructure, build community buy-in, improve the community's agrarian skill sets, and to take advantage of the new markets that will need to exist to maintain a local, sustainable food city (The Power of Community 2006).

Transition Towns

The example of Cuba transitioning through a fossil fuel shift demonstrates how the transition from fossil fuel dependence to a post–fossil fuel economy requires people to collaborate on solutions, create integrated systems for viable markets to work, and focus effort on planning for urban land to be dedicated to food growth. The security and resilience of the community food supply can be strong if planning and community involvement come together to share wisdom and to create a long-term vision for a community to thrive. It will take a new approach to planning through community engagement and building incremental successes, coupled with experimentation. The typical actors in sectors of agriculture, planning, water quality, parks and recreation, politics, and local government will need to learn how to integrate and innovate by coming together across sectors with open minds and through building trust. The growing transition town movement is a framework for moving through peak oil into a post–fossil fuel world in a way that builds the interrelationships necessary for us to get more involved while building healthy relationships with each other, our environment, the local economy, and our urban food system.

Transition towns represent a sustainable planning model based on Cuba's Special Period with the idea of implementing change before peak oil causes additional problems throughout the world. How does one move from a centralized, industrial, fossil fuel–based growth-at-all-costs culture to a sustainable, holistic, interdependent, and diverse culture? How does this cultural shift work simultaneously to create new systems for our energy and resource use, transportation, housing, and food? And how does this shift happen while the challenging realities of peak oil and decline of resources is happening? This is a very complex system and a process of engaging hugely diverse viewpoints and self-interests.

The transition timeline is a process that guides people through creating solutions together as a community for transitioning how they achieve their quality of life. It is a process that moves through building relationships, creating a vision together, and allowing people to self-determine what they would like to work on toward the vision that they have created for their community. The timeline, therefore, is a guide of a process for human change, including not being prescriptive and controlling, but based on relationships that can navigate complex futures. This is important for the flexibility necessary to work through the future that is impossible to predict, and to stay aware of innovative solutions that might not be seen when working toward an inflexible vision.

A unique, powerful strategy and tool that the transition town process employs is considering the stories we tell ourselves about what is or is not possible. The stories come from the media, our politicians, our economy, and what we learn from each other about how we create wealth and happiness, what is possible to do as community members, and what is possible for the future that we want to create. Community members in the Transition process work through identifying what are stories and what is reality. This allows people to work with the vision of the future and the challenges currently facing them. It is a story of working together to create a future that deals with the difficulties of peak oil versus doing nothing about it and hoping the solution emerges anyway, which is rarely realistic.

THE 12 STEPS IN THE TRANSITION PROCESS

#1. Set up a steering group an^d design its demise from the outset.
This stage puts a core team in place to drive the project forward during the initial phases.

#2. Raise awareness.
Build crucial networks and prepare the community in general for the launch of your transition initiative.

#3. Lay the foundations.
This stage is about networking with existing groups and activists.

#4. Organize a great unleashing.
This stage creates a memorable milestone to mark the project's "coming of age."

#5. Form subgroups.
Tapping into the collective genius of the community, for solutions that will form the backbone of the Energy Descent Action Plan.

#6. Use Open Space.
Open Space Technology has been found to be a highly effective approach to running meetings for Transition Town initiatives.

#7. Develop visible practical manifestations of the project.
It is essential that you avoid any sense that your project is just a talking shop where people sit around and draw up wish lists.

#8. Facilitate the great reskilling.
Give people a powerful realization of their own ability to solve problems, to achieve practical results, and to work cooperatively alongside other people.

#9. Build a bridge to local government.
Your Energy Descent Plan will not progress too far unless you have cultivated a positive and productive relationship with your local authority.

#10. Honor the elders.
Engage with those who directly remember the transition to the age of cheap oil.

#11. Let it go where it wants to go…
If you try and hold onto a rigid vision, it will begin to sap your energy and appear to stall.

#12. Create an Energy Descent Plan.
Each subgroup will have been focusing on practical actions to increase community resilience and reduce the carbon footprint.

Source: Chamberlin 2009

Figure 1.24 The collapse of Earth's resources is of great concern to human existence, and current agricultural practices are a contributing factor to the degradation.

Resource Collapse

Resource collapse encompasses all natural resources that are extracted, exploited, or processed. Resource collapse (Figure 1.24) is bigger than peak oil, and even bigger than the projected depletion of the world's natural gas, coal, and uranium. Many of our natural resources are limited and will eventually run out. Others require long time periods to replenish renewable materials. The dilemma we face is that with our current consumption practices, our natural resources are not being managed wisely and are disappearing at an alarming rapid rate, some of them never to return again. If countries fail to come to grips with these global challenges of resource collapse, the consequences will be felt worldwide via widespread famine, lack of raw materials, and industries unable to create goods and services that societies require to survive. One resource in particular that is being affected by poor agricultural practices is phosphorus. Phosphorus is a nutrient essential for plant life and found naturally in soil. It is removed from the soil by plants, and, in the case of agriculture, returned through fertilizers along with nitrogen and potassium. Most of the agricultural land in the world does not have enough phosphate. Yet phosphorus is vital if our ever-increasing global population is going to be fed. The consequence will be widespread famine if we do not meet this challenge, which is potentially more severe than a decline in oil.

Peak Soil

Not many are aware of the fact that soil stores approximately twice as much carbon as that in the atmosphere. The thin layer of soil that circles the planet's surface is approximately six inches deep and is the foundation of human civilization. This thin layer of planetary life supporting nutrients was formed over long stretches of geologic time as new soil exceeded the natural rate of erosion. Sometime in this century around the start of the Industrial Revolution, the world's soil erosion rate exceeded the world's soil formation rate.

Soil science is a technology that needs to be better understood if we are to focus on growing a sustainable future. Life-building soil is under attack by current land development and farming practices on a global scale. This accelerated erosion has potentially dire consequences for global food security. In his book *Dirt,* David Montgomery argues that soil erosion was the major cause of the fall of the Roman Empire, even though they did understand the importance of agriculture in their economy.

There are other examples in Montgomery's book of historic civilizations where bad practices resulted in a civilization's demise, and we can find fragments from their past and locations where the soil is now completely gone and cannot support a society anymore. There are places in northern Syria or the mountains of Ethiopia where famine is a constant fact of life. Years of overgrazing have destroyed the protective vegetation of the land. When the vegetation is gone, the soil is exposed to wind that removes soil and then turns the farmland into desert.

The lack of food security is growing wider globally because of this onslaught in the loss of the soil ecosystem. Wind and water take their toll, as do poor development and farming practices. These practices will increase soil degradation and deterioration at a more rapid rate than new soil formation can counter. Civilization on this planet depends on fertile soils. Ultimately, the health of the people cannot be separated from the health of the land. Conserving and rebuilding soils is a necessity we cannot live without in the rural and urban world.

The Loess Plateau—A Story of Degradation and Rehabilitation

American environmental filmmaker and sustainability advocate John Liu documents the world's ability to restore landscapes that have been degraded by climate change, poverty, and land mismanagement. Through film he leverages the power of the image to inspire hope and action. The issue to him is knowledge. For the public or for policy makers, ignorance should not an excuse. He has been documenting best practice methods for large-scale restoration of damaged or destroyed ecosystems.

One project in particular, The Loess Plateau Rehabilitation project, has given him a renewed sense of hope for the potential for poverty eradication through ecosystem rehabilitation as one of the ways forward to a more sustainable future. He was commissioned by the World Bank to document a project in the Loess plateau aimed at rehabilitating a barren and desolate area that was once highly fertile farm land. After centuries of deforestation and overgrazing, the region's ecosystems were destroyed. At a recent Bioneers conference I attended where Liu presented excerpts from the film, he described that his initial experience in arriving to the site was of skepticism as he stood looking at the barren wasteland. He described and then showed a place where the land was so scarred and was without even a hint of vegetation. The area did receive water as storms would deliver a river of water onto the land, but because of the lack of topsoil the water was immediately creating more erosion in its wake. The water did not infiltrate into the ground, as the soil was incapable of retaining any water.

Through successive trips, Liu documented the progress of the ecosystem rehabilitation program and began to notice small, incremental changes. The project followed the scientists and experts who drew up plans that focused on rezoning areas not acceptable for agriculture into conservation ecological zones to make them off limits to the local farming practice of using every piece of land available. To address the lack of water filtration, the environmental solutions focused on creating dams in existing gullies to hold water to create moisture in the soil. Combined with the ecological repair solutions, an equity-based educational program aimed at teaching better farming practices to the local inhabitants was instituted to engage them as part of the ecosystem repair. The goal was to help them expand their economic situation, currently at poverty level, by embracing the ecosystem rehabilitation project. Practices that were contributing to the land erosion, such as letting goats range free, were forbidden. The better farming practices required the local populace to abandon their destructive practices that only provided them with short-term marginal income. This was a hard requirement to accept for people who were already living in poverty. According to Liu, these new practices were initially met with much resistance, and at first he didn't think there would be much success in the efforts. He documents that the hardest practice for the locals to learn was in planting and protecting trees. Gradually, as they saw how the improvements began to change the land and to improve their ability to earn a living from it, they began to realize that if they worked with the ecosystem, it rewarded them.

Films such as Liu's *Hope in a Changing Climate* are potentially powerful tools that could be used to educate communities and policy makers about the seriousness of these issues by elevating the discussion on how ecology and culture can work together in solving it.

Environmental degradation issues also can be equated to urban and industrial areas of our cities where man has created "dead zones" where there is no biology. By taking away biodiversity, biomass, and organic matter in our cities temperatures will rise. By taking these away and replacing them with pavements, parking lots, and buildings as most cities do, the city loses the capacity to retain and infiltrate rainfall, which in turn creates more flooding. If cities, their buildings, and their systems—including transportation, water, waste, and food—were designed around these ecological principles, we would see a more sustainable and livable city in the future. One way to bring the ecological principles into the foreground is through urban agriculture landscapes.

We can find proof of the importance of properly caring for soil ecosystems from older civilizations that managed their land well and were able to maintain their land's fertility over long periods of time. Many of these examples have continued to flourish even until current times because of their good farming and development practices. Organic farming practices are a great model for handling degraded soils in our cities, but they also are a model for reducing pollutants and chemicals in our communities through our food. Better practices for soil management need to be incorporated into city policy. Policy makers can learn from urban ag landscapes the value of these practices.

Peak Water

Water is life. All humans have a cellular and biological response to water. Human beings are made up of 98 percent water. Humanity's deepest fears include the threat of too much water or having too little. Water has become the face of climate made visible.

> **A FEW WATER FACTS TO CONSIDER:**
>
> Lawn is largest crop in America; the average lawn receives 10,000 gallons of water per year.
>
> In single-family homes, 30 percent of all water is used for outdoor use.
>
> American households use approximately 100 gallons of water a day, while millions in the poorest countries subsist on less than five gallons a day.
>
> Women in developing countries walk an average of 3.7 miles to get water.

Due to population growth and associated land use impacts on water quality and quantity, water has become the limiting factor to growth. Water currency is becoming the new gold; however, because many nations currently pay very low prices for water, it is not obvious that the world is facing a water crisis of massive proportions. Consider what you paid for the last bottle of water you purchased. Less than 2.5 percent of the world's water is fresh, and only a fraction of that is accessible. In fact, 70 percent of the world's fresh water is locked up in ice. More than one billion people currently lack access to clean water. It has been predicted that by the year 2032, two-thirds of humanity may live in nations running short of water (United 2002, Freshwater 2010).

Flow by Irena Salina is an award-winning documentary investigation into the world water crisis. The movie presents a case against the growing privatization of the world's fresh water supply and illustrates how that supply is rapidly declining. It tells a story about the intersection of politics, pollution, human rights, and the emergence of what she describes as a powerful global water cartel through interviews with scientists and activists who shed light on the issue from both a global and human scale. Highlighting actual locations that examine the crisis firsthand, it dares to ask the question, "Can anyone really own the water?" This question is also being asked in states such as Utah, where state water policy dictates that you cannot collect rainwater on your own personal property. Rainwater harvesting is not allowed. How can water be owned when it runs through rivers and living bodies? Isn't the ultimate trust that every drop of rainfall will make its way from where it falls and travel across the land to the oceans and that the oceans will remain full?

The country of Ecuador has become the first nation on Earth to put the rights of nature into its constitution so that forests, lakes, and rivers are not just property but maintain their own rights to flourish. Any citizen may file a lawsuit on behalf of an injured watershed or destroyed forest canopy in recognition that its health is necessary to the common good. Some law schools in the United States are beginning to include the understanding and acknowledging of nature's rights into their curriculum.

The use of water in landscape design is elemental, with a history that stretches across cultures and back through centuries of constructed agriculture fields and public spaces. However, with today's increased emphasis on conservation and stewardship, water management must reflect a perspective of sustainability and resource management. More informed water management choices need to be based on specific climate and regional metrics and focused on increasing the health of the local water sheds.

Urban agriculture will potentially add to the pressure to our urban water systems and the growing global water crisis. When considering these landscapes, wise water management solutions must be balanced with practices such as rainwater harvesting, graywater use, and water-smart irrigation methods such as drip irrigation coupled with controllers based on up-to-the-minute climate data. Drip irrigation also improves water efficiency and effectiveness. Huge stormwater management infrastructure projects are very expensive. Therefore, an approach that incorporates more localized, decentralized stormwater approaches, including more green roofs and urban gardens, are less expensive ways to alleviate pressure on overstressed urban stormwater management systems. Through local organization and advocacy, communities can focus on changing zoning rules and create a national movement that makes urban farming and urban agricultural landscapes more acceptable and practical. The treatment and management of water needs to start with a bottom-up approach.

VIET VILLAGE, New Orleans, Lousiana

The Viet Village Cooperative Urban Farm Project (Figure 1.25) is a planned urban farm, community garden, and produce market in the New Orleans East neighborhood of New Orleans, Louisiana. The design was commissioned by the Mary Queen of Vietnam Community Development Corporation to serve the significant Vietnamese-American population of eastern New Orleans. Following a surge of immigration in the 1970s, the Vietnamese population established gardens, informal farms, and markets on a variety of sites throughout the community. This not only provided access to the traditional fruits, vegetables, and leafy greens of Vietnam but also allowed for the consumption of fresh, healthy, locally produced food, and for greater food security.

Figure 1.25 The Viet Village Cooperative Urban Farm incorporates an urban farm, community gardens, and produce markets into one cohesive design framework.

Sadly, this food network, containing over 30 acres of farmed land, was badly damaged by Hurricane Katrina in 2005. The Viet Village Urban Farm is the result of an effort to reestablish the network, while creating a permanent, robust, and highly designed facility for agriculture, animal husbandry, education, and produce distribution. Through a series of public meetings in the design process, the landscape architects have worked with community members and the development corporation to develop a plan that will meet the needs of the community and the site. The farm will be located on a 28-acre site in the center of the community. Its site design includes areas dedicated to plot-based family gardening (Figure 1.26), commercial plots for local restaurants and grocery stores, and areas for raising poultry and goats. The on-site market will provide space for individuals and families to sell their excess produce, stalls for local Vietnamese restaurants, and gathering space for Saturday markets and holidays expected to attract the larger Vietnamese community of Louisiana's coast (Figure 1.27). These various site elements are designed so that they can be phased independently, ensuring that some areas can be built as funding becomes available, even if the entire project is not underway.

Figure 1.26 Families can manage their own plots.

Figure 1.27 Viet Village is envisioned as a center for community activity beyond food.

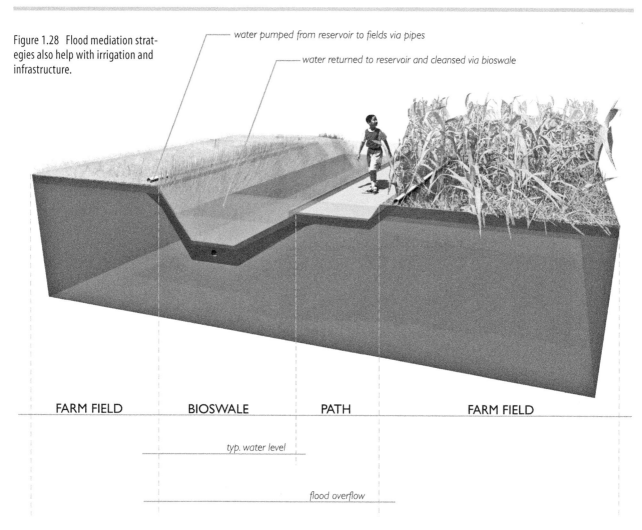

Figure 1.28 Flood mediation strategies also help with irrigation and infrastructure.

The group also wants to establish sustainable-based farming practices that will ensure the long-term health of the farm and community members. Toward this end, the farm will be completely organic and will incorporate practices such as integrated pest management, composting, crop rotation, cover cropping, and the use of alternative energy sources. The farm will also address on-site water management issues, as the site is nearly flat and consists of poorly draining soil. Irrigation water will be sourced from a central reservoir. The surrounding farm lands will be divided into watersheds, each of which uses bioswales to cleanse and convey irrigation runoff back to the reservoir (Figure 1.28). The watersheds can be activated one-by-one as different project areas come online.

Unfortunately, while the project has a strong vision and community support, the project has still not been realized due to politics and other challenges that have been thrown in its way.

Design Team:

Client:	Mary Queen of Vietnam Community Development Corporation
Landscape Architect:	Spackman Mossop Michaels
Collaborators:	Louisiana State University Urban Landscape Lab
	Tulane City Center
	University of Montana

Ecoliteracy and the Current Unsustainable Urban Food System Model

Ecoliteracy is the understanding of the connection and relationship between ecological health and human health. To be ecoliterate is to understand the principles of ecosystems and use those principles for creating sustainable communities. The term was coined by American educator David Orr and physicist Fritjof Capra in the 1990's and added a new value to education that is concerned with the health of people and the planet. With an understanding of ecological literacy, perceptions shift to an understanding that protecting ecosystems is a basic principle for prioritizing thoughts and actions in a sustainable society that is aware of the importance of living within the ecological capacity of the earth. The current urban food system model does not foster connections between food and the health of communities or ecosystems. Industrial agriculture plays a large part in the destruction of ecosystems. The current industrial food system is not a sustainable model. It fosters:

- An urban planning infrastructure based on separation rather than integration
- Food security and national health issues
- A commodities subsidized culture
- Food injustice
- Processed and nonnutritious food over organic nutritious food
- A disconnect with real food in urban areas
- Soil and mineral depletion
- Pollution of toxic fertilizers and chemicals into to local watersheds and soil sheds
- A decline in ecoliteracy

The interrelationships between food, health, and the environment have never been a more important topic. Sustained life is the property of an ecosystem. The earth has 2 billion years of ecological experience in sustaining life on the planet. Planning for our food system also needs to become part of this ecological model. "A trend has been unearthed regarding the ecological literacy of university students entering into a bachelor of education program. An analysis of the meaning contained in participant definitions has revealed that the vast majority of teacher candidates, graduates of many different universities, are unable to explain the meaning of key integrating ecological concepts at even a minimal level of maturity, alluding to a possible systemic problem" (Puk 2012).

David W. Orr, who is the Paul Sears Distinguished Professor of Environmental Studies and Politics at Oberlin College, points out that we spend 12 to 20 years educating ourselves on a part of the whole, a discipline or subdiscipline of specialization, without any training of the integration and awareness of the whole and how it is unified (Orr 1991). We can see this in the silos that form between disciplines in our economic markets, government, and decision making. Our lack of ecological literacy gives us little chance to think systemically about the food system we create and encourage with our purchasing. It is part of why we do not account for more than the single bottom line of the dollar when producing, selling, and purchasing food products. It is difficult without ecological education to understand the values of the ecosystem services that are produced or destroyed by our actions. Even more difficult is trying to understand how to create food systems that have regenerative feedback loops that increase the ecosystem services we need for a robust urban ecology and high quality of life. Not many of us are taught how to systems-think the benefits provided to salmon when upstream people are seed saving and managing the soil biology in our community gardens.

Ecosystem Services, Systems Analysis, and Metrics

The integration of sustainable technologies with system metrics presents a potential path for balancing the economic, social, and ecological choices that urban agriculture landscapes require. Considerations for combating childhood obesity, fostering community interaction for multiple age groups, and fostering biodiversity represent just a few of the social metrics for human health and well-being that can add intangible value to the urban environment.

When working toward developing ecosystem services, we need to also remember that our understanding of the current state of our ecosystems is limited. A focus on a widespread ecoliteracy program has been shown to be necessary as recent studies have demonstrated that in inner city schools as well as affluent ones, a majority of students have not been able to identify naturally growing fruit or vegetables and believe most food just comes from a store. Urban agriculture can become a big part of the solution for educating communities on the connections between food and health. Designers, planners, and policy makers can help by creating the opportunity for communities to start taking their health into their own hands incrementally through healthy soil management and food production to begin the reduction of our dependence on products that include the creation of toxic side effects from their use.

Ecosystem services are those ecological systems and benefits (Figure 1.29) that nature provides naturally that support human life. These include clean air, habitat corridors creation, stormwater treatment through watersheds, soil health that increases food vitality and protects from erosion, biodiversity, carbon sequestration and air quality, climate comfort control, reduced heat islands, connections of people to nature, and many more:

- Ecological, sociological, and economic metrics are all necessary to map and consider when looking at an approach that values ecosystems services. New rating systems such as the Sustainable Sites Initiative, SITES, which is a partnership of the American Society of Landscape Architects (ASLA), the Lady Bird Johnson Wildflower Center of The University of Texas at Austin, and the United States Botanic Garden. SITES was created in 2005 to fill a critical need for guidelines and recognition of green landscapes based on their planning, design, construction, and maintenance. The voluntary, national rating system and set of performance benchmarks applies to sustainable landscapes in areas with or without buildings (Calkins 2012).
- Through ecological-based practices including plant choices, sustainable installation techniques, and organic amendments such as composting, the soil can be revitalized. This will add to food vitality, make food safer, and also reduce erosion in urban environments. Erosion is a leading factor in increased city flooding. Metrics that measure these impacts provide feedback to adjusting or enhancing choices for maximum benefit.

34 Designing Urban Agriculture

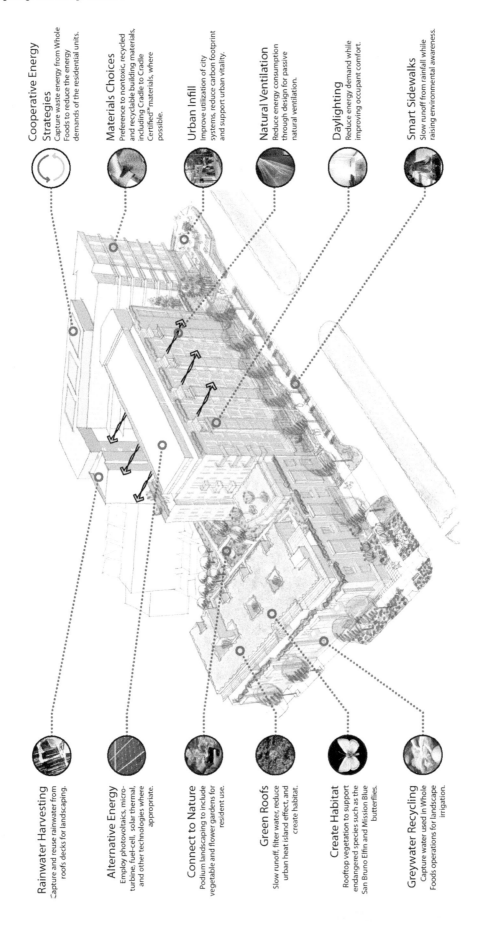

Figure 1.29 Good design exploits the services nature already provides to reduce the resource consumption of a project.

- Through the use of metrics we would be able to provide data that support designing alternative energy sources of the future using vegetative biomass to produce heat through thermal conversion or biogas digestion, alga culture, or anaerobic digestion. Currently, these sources are not yet an economical alternative.
- Biodiversity is the variety of life. By restoring native habitat, removing invasive species, adding plants that attract beneficial habitat and pollinators, the caring for, planning, and monitoring of urban ag landscapes can add to the increased biodiversity in the city. Metrics for measuring these impacts would be beneficial to providing guidelines for developments to follow to increase biodiversity.
- Trees and plant biomass add to the city's ability to sequester carbon and eliminate pollutants from the air. Cleaner air reduces risks to health problems such as asthma, which have increased tremendously in the past decade. Metrics would guide our choices for designing solutions that increase the health of our communities. Trees also have the ability to provide for increased climate control and reduction of the heat island effect generated by cities.

An example of the disruption of an ecosystem service is habitat loss. Habitat loss is the most widespread cause of species endangerment in the United States, affecting 85 percent of imperiled species (Wilcove & Master 2008, p. 416):

- Pollution causes many species to become endangered, especially a large proportion of aquatic life.
- Estimated damage and control cost of invasive species in the United States alone amount to more than $138 billion annually (Pimentel 2005).

One example pertains to the king crab: It is too soon to estimate the potential damage that might be caused by the migration of the king crab population in Antarctic waters to areas where there has not been a predator in over 40 million years. Due to the trend in warming waters, even by one degree, an underwater habitat of diverse sea life will become prey to these crabs as they move into new territories that used to be too cold for them. Sea creatures that did not have a predator to worry about will now begin to fall prey to the invading crabs as they expand their territory. This is one example of how just a minor shift caused by climate change may prove to have wide-ranging impacts on the natural environment. Many pest issues can be resolved with strategies that foster habitat creation and therefore the welcoming of a balanced and self-maintained ecologically sound landscape. When changes occur based on larger aspects of climate change and shifting populations, stronger objectives will need to be put in place to achieve these goals.

Systems-Based Analysis

The principles for integrated systems are to interconnect the systems so that the output of one system is the input for another system and treat all waste materials of the system as resources; be able to generate value and revenue from these resources; achieve multiple benefits from each system where possible; and coordinate the location of resource producers and the users near enough to each other to facilitate resource exchange. These actions all increase the ability to achieve the goal of zero net waste.

Setting the Stage for a Sustainable Food System

A sustainable urban food system is an integrated approach to bring food back into the city based on the integration of people, their living environments, and food. It requires a system-based approach that spans the full spectrum from urban to rural to wild. It takes urban agriculture to a planning and design level, making it a more fully integral part of the city—not separate as an industrial-oriented land use, but rather a way to integrate biodiversity, agriculture, social/cultural/community, and economics on a daily and more personal level. This integration of natural systems with urban systems into city infrastructure needs to include a sustainable urban food system.

Integrated systems thinking must focus on solutions based on the interconnectedness of the systems as a whole unit rather than separate independent units. The network of the systems is a vital component that functions better with integration, not separation. It requires that urban agriculture and the development and viability of a city's food shed be incorporated into the planning process equal to the viability of water sheds and transportation networks. As we learned earlier, if food is the "gateway drug" to sustainability, we could access these markets through kids in schools first, then parents. Local, healthy food creates positive externalities.

One way to describe systems is via the *Cynefin Framework,* which outlines four types of systems and how to behave in them to make change happen—simple, complicated, complex, and chaotic. Simple systems have a connection between cause and effect that is obvious to all: sense—categorize—respond. Complicated systems use expert knowledge and analysis to identify and act on the clear cause and effect relationships: sense—analyze—respond. In complex systems, the relationship between cause and effect can only be seen in retrospect of acting; you probe—sense—respond. And chaotic systems have no relationship between cause and effect: act—sense—respond. Planning, signing, engaging, and executing a new urban food system is a complex system with complicated systems within it (Snowden 2007).

It is important to understand how change happens in different systems when you are a designer change agent. Each system needs a different type of leadership. If a soil test contains toxins, it needs to be cleaned up; simple. How to design the process and techniques and products of the toxin cleanup process requires an expert to design, manage, inspect, and test. In order to help the community, owner, politicians, and other stakeholders of that property to invest in the value of clean soil is a complex system. How the project lead can work with everyone to make the project happen cannot be predicted and only understood after the project is complete.

The number of factors and the dynamic relationships between those factors are too complex for any simple predictions. A designer change agent has to lead this process by probing the system, asking questions, making proposals, trying tests, creating incremental successes, and facilitating group processes with the stakeholders. Each step requires an incremental step, observation of the changes to the stakeholder group to learn from, then adjusting and moving forward with another incremental step. This is change management of complex systems.

Change management of complex systems requires numerous skill sets in addition to community input, planning, design, and project management in order for the stakeholders to stay engaged with the process even when the outcomes are not specifically predictable and they are based on relationships to get accomplished. When forming your team and partnerships, look for expertise in fair decision-making processes, social technologies such as World Café and Open Space, collaborative work, team dynamics and managing diversity, community engagement, personal awareness, and managing through fear of change and communications. An important lens through which to look at these skill sets is the philosophy that change should happen *with* the community, not *to* the community. This ensures that the change management is always considering the buy-in and adoption of any change by the community to ensure its sustainability.

John P. Kotter's Eight Steps to Successful Change

American John P. Kotter (b 1947) is a Harvard Business School professor and leading thinker and author on organizational change management. Kotter's highly regarded books *Leading Change* (1995) and the follow-up *The Heart of Change* (2002) describe a helpful model for understanding and managing change. Each stage acknowledges a key principle identified by Kotter relating to people's response and approach to change, in which people *see, feel,* and then *change*.

Kotter's eight-step change model can be summarized as follows:
1. **Increase urgency:** Inspire people to move, make objectives real and relevant.
2. **Build the guiding team:** Get the right people in place with the right emotional commitment, and the right mix of skills and levels.
3. **Get the vision right.** Get the team to establish a simple vision and strategy; focus on emotional and creative aspects necessary to drive service and efficiency.
4. **Communicate for buy-in.** Involve as many people as possible, communicate the essentials, simply, and appeal and respond to people's needs. Declutter communications—make technology work for you rather than against.
5. **Empower action:** Remove obstacles, enable constructive feedback, and get lots of support from leaders—reward and recognize progress and achievements.
6. **Create short-term wins:** Set aims that are easy to achieve—in bite-size chunks. Manageable numbers of initiatives. Finish current stages before starting new ones.
7. **Don't let up:** Foster and encourage determination and persistence—ongoing change—encourage ongoing progress reporting—highlight achieved and future milestones.
8. **Make change stick:** Reinforce the value of successful change via recruitment, promotion, and new change leaders. Weave change into culture.

Kotter's eight-step model is explained more fully at kotterinternational.com.

Scale Aggregation

Urban agriculture's potential to address the challenges of our food system remains unknown. Although the popularity and trendiness it is experiencing can be a big boon to the creation of local businesses and entrepreneurs, urban farms have not yet found a way to thrive in a mass market economy. Most still heavily rely on volunteer labor and grant funding and operate on donations rather than function as profit centers that add value to local economies. They may be at the forefront of ecological sustainability, but economic sustainability eludes them at a scale that would begin to make them an enticing market proposition. That is a serious problem because they are unlikely to fulfill their aspirations and make a meaningful dent in the problem of food insecurity if they are forever running on the roller coaster of private foundation funding or public government and institutional grants.

Scale is one of the larger issues that urban agriculture must confront if it will succeed in market economies and satisfy the increasing demand for sustainable food. Approximately 60 to 70 percent of food dollars are being spent outside the community in many food-insecure neighborhoods. Urban farms are only closing a fraction of that gap at approximately 10 percent, according to Brahm Ahmadi, the cofounder of People's Grocery in Oakland, Ca. So Brahm and other food security experts predict that we will need to look at how we can scale urban agriculture differently to make a more significant impact.

One new way a few food and urban farm entrepreneurs have begun to test scaling differently is with the notion of scale aggregation. Scale aggregation is a land management technique that consolidates numerous smaller farms in urban environments to create viable economic return into citywide networks.

Big City Farms, Baltimore, Maryland

Big City Farms is an urban agricultural business (Figure 1.30) that aims to grow into a large network of for-profit, organic, urban farms and farmers, making productive use of underutilized land across the city. Founded in Baltimore, Maryland, by businessman Ted Rouse, the company operates its own pilot farm, which it uses as a testing ground for construction and growing techniques.

The Big City Farms pilot farm, built in February 2011, consists of six 3,000-square-foot plastic hoop houses (Figure 1.31) on the site of the city's former maintenance garage and parking lot. The site is considered a contaminated brownfield, and as such has limited usage potential for more traditional urban land use. The farm uses imported organic soil, however, and is able to operate on top of the preexisting pavement. The farm primarily grows leafy greens, such as lettuce, fennel, and Swiss chard, which are sold to local restaurants, markets, distributors, and individuals.

The lessons learned at the pilot farm will be put to use by new farmers who join the network, but Big City Farms will provide more than just experience and knowledge to those who join their business. The company will sell plant plugs, compost, and enriched growing medium, and will provide training in construction and operation of a hoop house farm. It will also handle collection, processing, and distribution of its members' crops, plus legal issues, marketing, and sales, reducing the business risks for its farmers. BCF also hopes to provide financing for its members at a later stage.

Food Cities: Ecology + Urban Agriculture 39

Figure 1.30 Big City Farms started with six large-scale hoop houses in Baltimore.

Figure 1.31 The hoop houses are located on a brownfield site so the farm imports organic soil.

As a benefit corporation, Big City Farms has a "triple bottom line," which pursues social and environmental goals in addition to profit. One primary social aim is to generate green jobs at a time of high unemployment. It estimates that there are over 1,000 acres of underutilized land with limited development potential in the city, each acre of which could provide 10 jobs if farmed using the Big City Farms system. The community should benefit as well from the fresh, healthy, organic, and nutrient-dense produce, available at competitive prices. Since BCF sells its produce within 24 hours of harvesting, varieties can be selected for their nutritional content and taste rather than their durability.

To address environmental impact, the Big City Farms growing system uses organic methods to reduce contamination by traditional fertilizers, while the hoop houses take advantage of solar heating to allow growing throughout the year. Perhaps more importantly, Big City Farms' produce is grown and distributed locally, drastically reducing the amount of fossil fuel used to get products in the consumer's hand. Similarly, the BCF will use a short supply chain to provide its growers with the materials needed.

Developer/Entrepreneur: Tim Rouse

Resources

Calkins, Meg. *The Sustainable Sites Handbook*. Hoboken: John Wiley & Sons, Inc., 2012.

Centers for Disease Control and Prevention. "Childhood Obesity Facts." 2009. www.cdc.gov/healthyyouth/obesity/facts.htm

Chamberlin, Shaun. *Transition Timeline: For a Local, Resilient Future*. White River Junction: Chelsea Green Publishing Company, 2009.

"Freshwater Crisis." *National Geographic*. (April 2012)

Montgomery, David. *Dirt: The Erosion of Civilizations*. Berkeley: University of California Press, 2012.

Orr, David. "What is Education For?: Six myths about the foundations of modern education, and six new principles to replace them." *The Learning Revolution* Winter 1991(1991): 52.

Pimentel, David, Rodolfo Zuniga, and Doug Morrison. "Update on the environmental and economic costs associated with alien-invasive species in the United States." *Ecological Economics,* Volume 52, Issue 3 (2005): 273–288.

Puk, Thomas G. and Adam Stibbards. "Systemic Ecological Illiteracy? Shedding Light on meaning as an Act of Thought in Higher Learning" *Environmental Education Research*, v18 (2012).

The Power of Community: How Cuba Survived Peak Oil. DVD. Directed by Faith Morgan. Yellow Springs: The Community Solution, 2006.

Snowden, David. J., and Mary E. Boone. "A Leader's Framework for Decision Making." *Harvard Business Review* (November 2007): 69–76.

Standage, Tom. *An Edible History of Humanity*. New York: Walker Publishing Company, Inc., 2009.

Steingraber, Sandra. *Living Downstream: An Ecologist's Personal Investigation of Cancer and the Environment*. Philadelphia: Da Capo Press, 2010.

United Nations Environment Programme. *Global Environment Outlook 3*. 2002. www.unep.org/geo/GEO3/english/pdfs/chapter2-5_Freshwater.pdf

United States Department of Agriculture. *Dietary Guidelines for Americans, 2010*. www.cnpp.usda.gov/dgas2010-policydocument.htm

CHAPTER 2
Planning Strategies for Urban Food Systems

Prairie Crossing, Greyslake, Illinois, USA

Prairie Crossing, a planned "village" neighborhood in Grayslake, Illinois, is a self-described "conservation community" (Figure 2.1). Conservation has been central to the community's design, vision, goals, and activities since its inception. A central component to their conservation strategy is the 100-acre, certified organic, peri-urban farm on the neighborhood's western border (Figure 2.2). The farm enables Prairie Crossing's residents to support hyper-local agriculture through a variety of initiatives and businesses.

In 1986, a developer proposed that a traditional suburban neighborhood be built on the land that the neighborhood currently occupies. Prairie Crossing was first envisioned by Gaylord Donnelly, a conservationist and chairman of RR Donnelly, a large Chicago-based printing company, who opposed the plan and organized a group of other area property owners to purchase the land by forming the Prairie Holdings Corporation to secure the land for an environmentally friendly development. When Donnelly died in 1992, his nephew George Ranney took over the cause.

Figure 2.1 An aerial view of the Prairie Crossing Conservation Community.

Figure 2.2 The hundred-acre farm is on the neighborhood's western border.

George, and his wife, Vicki, saw an opportunity to create an environmentally sound neighborhood while simultaneously protecting much of the surrounding habitat. They perceived that typical suburban developments sever connection to the landscape. The Ranneys and their partners planned for the inclusion of a farm in the neighborhood from the very beginning. The visionary leading the planning effort for the conservation community master plan was renowned landscape architect William "Bill" Johnson. Though many firms have been involved though the years, it is Bill who set up the overall site planning strategies for the community. Bill was instrumental in laying out the master plan to be mindful of view sheds that were important because of the flat prairie site. The agricultural component, the farm, was also an important concept from the start because of the regional farm vernacular and as part of the identity and structure in creating a self-sufficient community. Sited on the edge of the community, it also serves to buffer the community from the adjacent landfill area.

The vision for the farm was expanded by Mike Sands, who was brought in as the corporation's environmental team leader. He foresaw that the farm would have a greater chance at success if the residents were more than just customers, but were connected to and intimately involved with the farm's endeavors (Figure 2.3). The farm was one of the first components of the neighborhood to be built, and the first crops were harvested in 1994, a full year before the first homes were completed. The team included Steve Applebaum, an ecologist who was instrumental in setting up the wetland and prairie ecologies of the surrounding open space conservation areas. Bill Johnson also worked with Peter Shaudt's office as the local landscape architect to design all of the public spaces, house prototype landscapes, and weave the public areas together. The majority of the landscape is based on the prairie ecology, as part of the sustainability plan. All in all, about 30 to 40 consultants played a role in creating the lifestyle community.

One of the main undertakings at the farm is Prairie Crossing's own Farm Business Development Center. The administration and residents at Prairie Crossing recognized a need for more environmentally conscious farmers and farms to feed the region's growing population. Toward this end, the center serves as an incubator, providing training, land, material resources, and financing to help educate and enable a new generation of farmers. These upstart businesses "graduate" from the program within five years, after which some elect to lease land from the farm and, in turn, provide mentoring to the next generation.

Another important component is the Prairie Crossing Learning Farm, with its mission "to educate and inspire people to value healthy food, land and community through experiences on our farm" (Prairie 2012) (Figure 2.4). The three-acre farm contains greenhouses, an organic

Figure 2.3 Residents of the community working and learning on the farm.

free-range hen house and poultry pasture, and a fruit orchard, all of which are used as learning environments for their Summer Farm Camp and After-School Farm Camp. The Learning Farm also hosts the Prairie Farm Corps, which provides paid job training to Lake County teenagers. Participants learn about all aspects of a farm operation, including cultivation, maintenance, marketing, sales, and even cooking (Figure 2.5). The students sell their produce through the Prairie Farm Corps CSA, the proceeds of which help fund the learning farm.

The farm at Prairie Crossing is also home to several for-profit agricultural businesses, such as the 40-acre Sandhill Organics. Though they are private endeavors that lease the land and sell to both residents and nonresidents, these businesses also serve as mentors for the Farm Business Development Center and the Learning Farm. Both Mike Sands and Peter Schaudt believe that the Ranneys, with the help of the multidisciplinary design team, have been successful in selling the conservation community as a way of life. Both concur that this type of community that is integrated with an open space habitat system—including the farm works at a certain scale and Prairie Crossing at 700 acres—was a good place to start. Of all the lessons learned, Mike said that he would like to see the farm located closer to the town's center rather than being located on the edge of the community. Placing the farm as a main focal point of the community would provide an opportunity to enhance the relationships and further the discussion on the connections between nature, community, and health.

Figure 2.4 The farm is an integral part of the community's productive open space infrastructure.

Figure 2.5 Seminars in farm operations are offered to the residents. Community education is a primary focus of the learning center.

Partial list of Planning/Design Team:

Planner:	Bill Johnson
Environmental Team Leader:	Mike Sands
Ecological Consultant:	Steve Applebaum
Landscape Architect:	Peter Lindsay Schaudt
Planner:	Peter Calthorpe
Planner:	Phil Enquist, SOM

PART 1
BASIC FUNDAMENTALS FOR URBAN AGRICULTURE + ECOLOGY

In recent years, there has been a tremendous upsurge of interest in growing food in urban environments. This has primarily been in response to concerns about food prices, food miles, and the negative impact of current industrial agriculture practices on the environment. It has also been because people want better access to good, healthy, and affordable food, and to enjoy cultivating beautiful green spaces and meeting local people in their own community. According to Josh Viertel, former president of Slow Food USA in a recent interview there are now millions of people planting urban farms in the United States, which has been recorded as a 35 percent increase in just one year. 2010 surveys indicate that over 15 percent of the US population now has a backyard garden (American 2010). The emphasis has also been on how these landscapes are built on more sustainable goals and values. This chapter will discuss the types of methodologies, tools, and planning strategies that aid in designing these new urban agriculture landscapes to meet the rising demand and to look at their capacity to expand the idea of the sustainable city

The planning process for an urban agriculture system is currently overly complex simply because it is new to the project development process for most stakeholders in both private and public sectors and because many ordinances and laws exist that limit its potential as a viable ecological, cultural, and market-economy-driven system in the city. And, it really is a complex system of connections that are in need of being harnessed or linked in order to create thriving urban ag enterprises and successful, productive landscapes. Making decisions about urban food cities needs to be done with the community members who will be directly involved in making sure the food system that is set up by those decisions carries on. The decision makers are likely not the stakeholders who will be responsible for the ownership and care of the food system that is set up on the land the planning team designates for food use. In order for a community to support the food system they will have to have a stake in its success, a sense of ownership and pride, and care for the work and value that it can provide to the local quality of life. In order to execute the creation and sustainability of this new more integrated local food system network within the city the community engagement process is a key factor to developing the ownership and long-term buy-in that is needed for it to be accepted, embraced, and sustained by the community.

In Part 1 we will begin by exploring key definitions and vocabulary being used by designers, academia, farmers, and the public to describe urban agriculture and their landscape design elements. In Part 2 we will move on to discussions about what an integrated systems approach might look like and the criteria required to design for integrated systems.

Defining Urban Agriculture

In its broadest sense, urban agriculture encompasses the promotion of food, including meat and fish, and the practices of cultivating, processing, and distributing that food *in* (urban) or *around* (peri-urban) the city (Figure 2.6). The relationship of the urban food system to the peri-urban food system is an important one that adds to opportunities for integrated resource management. It adds directly to the energy capacity of the systems and increases the ability to create a district wide network that connects the system flows into and out of the city in a cyclic process.

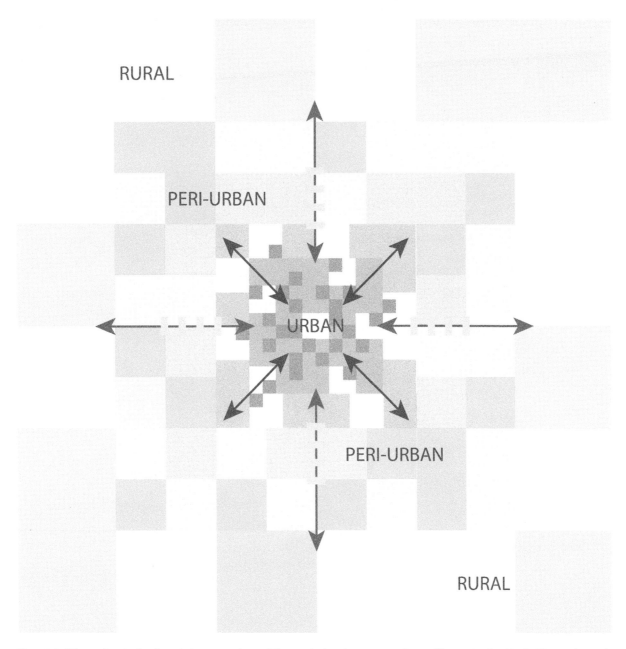

Figure 2.6 Urban and peri-urban boundaries are starting to shift towards the urban center as the need for growing local food addresses the needs of neighborhoods for food security, food justice, and community health.

In the past few years, media images for urban agriculture landscape have focused primarily on nonprofit-based urban city farms and community gardens as the main visual iconography but the reality is that urban agriculture is much more encompassing in diversity than limiting it to only land set aside to grow crops to harvest, share, or sell. The food system (Figure 2.7), when broken down into a series of its eight basic components, is in actuality more of a cycle and inclusive of growing and production, harvesting and processing, retail and markets, eating and celebrating, distribution and storage, waste and nutrient recycling, food education and outreach, and tying into policy and advocacy. Food security and food justice are drivers of the systems in areas where the systems are broken down or lacking in the community (de la Salle 2010).

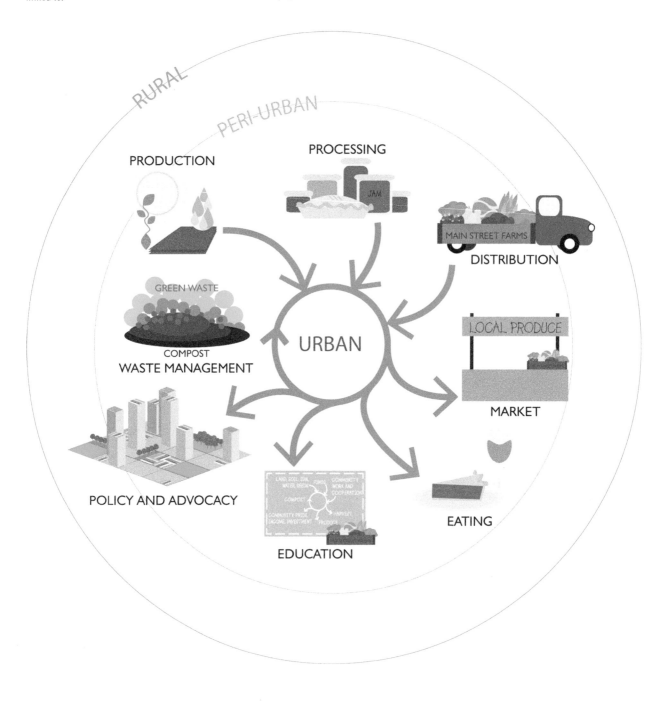

Figure 2.7 A food system is a cyclical process consisting of eight basic components. A city would link the food system of each neighborhood to the citywide food system they are linked to.

Figure 2.8 Young boy eating his first fresh-off-the-vine tomato in the Miller Creek Middle School edible garden during an evening family harvest event held weekly in the summertime.

City planners should address the citywide food system by first diagramming the food sheds of each city neighborhood and then connecting them to the overall citywide food system. The dilemma with explaining food as a system within the city seems to be limited by our inability as a society to think of food in this manner. Cities tend to view urban agriculture as a temporary activity, not a permanent one. In the past few decades, urban agriculture has only been thought of as farmland and has been related to rural areas outside of the city limits. People are so disconnected from thinking about the food system as something they are a part of that this becomes the first educational hurdle to tackle. Getting someone to taste food that comes from their own garden that they grew is a first step toward this realization. Anytime you are able to get someone to visit a farm and taste the food right off the plant, the ability to start a more real conversation about food has begun. It's even more rewarding with a classroom of children, especially if they have never eaten some of the vegetables or herbs you might get them to taste (Figure 2.8). These types of local food experiences begin to slowly change the cultural food beliefs and expand the definition in society to embrace food as part of the community infrastructure systems. If we look at food in this manner, we can begin to reduce the amount of people experiencing the effects of a food dessert and increase the ability to foster a more healthy community. The net effect will be to build a more health-conscious society that values healthy living as a natural extension of the services a city must provide to city dwellers.

Urban agriculture is moving from just a practice for earning an income and small food-producing activities to a more sustainable practice that focuses on promoting local food production as an energy-saving resource that is central to creating vital urban communities. It needs to become even more central to city planning as food security and food safety become issues that cities need to address along with the increase in population that is creating a strain on a global level with regards to food availability and health. In current practice, the term *urban agriculture* does not necessarily mean that food production itself is based on a sustainable methodology or procedure but when combined with an ecological-based approach it does. With the recognition of natural resource decline and the advance of environmental degradation in cities today, urban agriculture is taking on new meaning in bringing ecological-based systems back into the city as a vital part of the solution to creating more sustainable cities. This does require a paradigm shift in thinking about food as an integral part of the city's framework.

Traditional versus Environmental

Let's look at two different definitions of urban agriculture to see how the basic traditional viewpoints and the more ecological-based viewpoints are beginning to define the changing awareness of urban agriculture as part of a sustainable system in a city. The differences between an integrated systems approach versus a nonintegrated systems approach are important to understand when designing urban ag landscapes.

Traditional Definition

The Food and Agriculture Organization of the United Nations (FAO) definition of urban agriculture uses the broadest terminology and thus does not limit it to a sustainable practice:

> An industry that produces, processes, and markets food and fuel, largely in response to the daily demand of consumers within a town, city, or metropolis, on land and water dispersed throughout the urban and peri-urban area, applying intensive production methods, using and reusing natural resources and urban wastes to yield a diversity of crops and livestock (Smit, 2001, 1).

This definition, while it acknowledges meeting the nutrient needs of a city and seen as being a resource for the city's economic benefit, does not reconcile aspects of food security and community health, or allow for organizations that might provide some of these benefits as part of the urban agriculture system. It also does not address the aspect and benefits for sustainable agriculture or the role of agriculture as it relates to ecosystem health.

Environmental Definition

The Council on Agriculture, Science and Technology (CAST) defines urban agriculture to include aspects of environmental health, remediation, and recreation:

> Urban agriculture is a complex system encompassing a spectrum of interests, from a traditional core of activities associated with the production, processing, marketing, distribution, and consumption, to a multiplicity of other benefits and services that are less widely acknowledged and documented. These include recreation and leisure; economic vitality and business entrepreneurship; individual health and well-being; community health and wellbeing; landscape beautification; and environmental restoration and remediation (Butler and Moronek 2002).

Planners, designers, and environmentalists are more responsive to this definition since it addresses issues of sustainable design and ecosystem services, and the health and vitality of a community. This definition will be the one that this book considers the baseline for urban agriculture landscapes. When in doubt of the sustainable nature of the word being used within this book, there will be an emphasis placed on ecology—thus, urban agriculture plus ecology.

Urban Agriculture Landscapes

Urban agriculture landscapes are any landscapes that promote the integration of people, their living environments, and food. These productive landscapes are aiding the promotion of a sustainable urban food system approach. They include food landscapes growing in an urban or peri-urban environment, whether on an income-generating scale or not. An urban agriculture landscape may include food-producing plants and animals or both. It may or may not be based on sustainable methodologies or procedures unless that is part of the landscape's mission and goals, but more and more urban agriculture landscapes are organic or practice ecological best practices that remediate and heal the urban environments they are located in.

When urban agriculture is thought of as part of the green infrastructure of the city, promoting ecological biodiversity and social sustainability, then it is elevated to a sustainable system no matter what the scale, type, or location of the landscape. Examples of urban ag landscapes are diverse and range in size and complexity. An urban agriculture landscape could be a city street planted as an orchard with fruit trees that can be harvested by the local neighborhood or as small as a window box planter in front of a restaurant used by a chef in the daily menu. It could be a 10,000-square-foot community garden on a city lot or it could be a 3,000-square-foot edible garden terrace in an apartment high-rise that includes herbs and vegetables for tenants to grow, harvest, and eat. It could be a one-acre rooftop farm that grows, harvests, sells its produce, educates the community, and trains local youths for green jobs. Or it could be an edible school garden in planters on an asphalt lot that teaches students about nutrition and real food or a vertical garden grown on a blank building façade that has herbs and seasonal fragrance that pedestrians just enjoy as they walk by. It could be a corner plaza with perennial herbs and medicinal plants that citizens and passersby could pick for their own use. It could be a victory garden at the city hall grown for any citizen volunteer to tend and or harvest. The "it could be" list is really as unlimited as there are places and people in the world to think of them.

A few key urban agriculture definitions:

Food landscapes, edible landscapes, productive landscapes, and urban agriculture landscapes: These terms as used in this book are synonymous and are used interchangeably to describe landscapes that promote the integration of people, their living environments, and food. These landscapes are based on an integrated sustainable urban food system approach that fosters the health of the ecology, community, and economic vitality of a city with an ultimate goal of sustainable resiliency. This approach is rooted in environmental ecoliteracy, community interaction, and the integration of a citywide sustainable infrastructure system network. Landscapes can encompass the scale of a city as a landscape or the smaller scale of specific landscapes within the city.

Edible urbanism: The term *edible urbanism* is not an officially coined word for urban agriculture. It could be used to describe planning methodologies for incorporating food sheds and their associated system connections into the infrastructure systems of a city.

Agtivist: People who champion urban agriculture in their cities and towns, many of whom are focused on advocacy, changing policies, grassroots action, and the locavore food movements. Access to real, healthy food, farmers markets, food

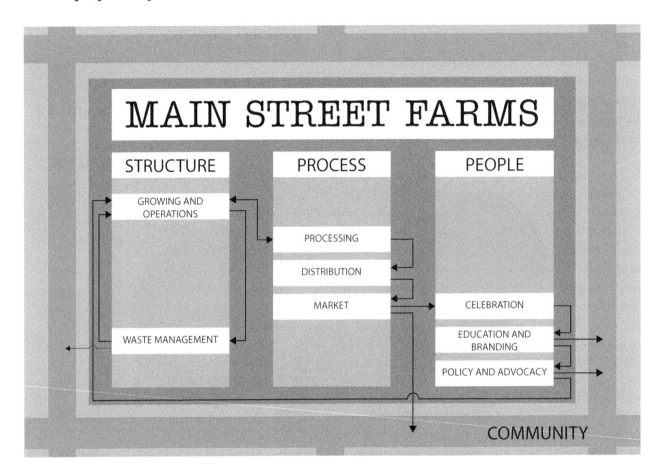

Figure 2.12 In a nonnatural system, some parts can be dependent on another and others can be independent, resulting in breaks in connections unless they are consciously connected.

Two solid sources for understanding a more comprehensive view of systems thinking and systems-based approaches are *Thinking in Systems* by Donella Meadows and *Steps to an Ecology of Mind* by Gregory Bateson. Donella Meadows was a pioneering American environmental scientist, teacher, and writer. She was co-author along with her husband, Dennis Meadows, on *Limits to Growth*, first published in 1972. They also co-founded an international network for leading researchers on resource use, environmental conservation, systems modeling, and sustainability that meets every year to advance critical global thinking on issues. *Thinking in Systems* is a great book that describes how systems work in a way that nonscientists can easily understand yet complex and rich in detail to engage scientists as well. The subject matter has been simplified so the average reader can better comprehend it. Gregory Bateson was also another proponent of systems thinking and the intersection of cross-disciplinary thinking. He saw systems as a more holistic means of planning our cities as the way to address and problem solve global environmental issues and other complex planning issues that cities of today are facing. This book is rather dense for the average reader but fascinating in its diversity of ideas and thought proposals.

Ecoliteracy and Systems Thinking

Ecoliteracy, also referred to as ecological literacy, is the ability to understand the role that natural systems play in order to sustain life on the planet and that humans are part of

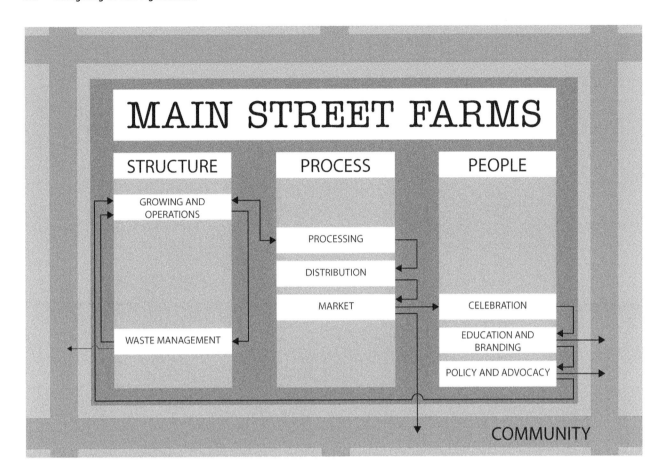

Figure 2.12 In a nonnatural system, some parts can be dependent on another and others can be independent, resulting in breaks in connections unless they are consciously connected.

Two solid sources for understanding a more comprehensive view of systems thinking and systems-based approaches are *Thinking in Systems* by Donella Meadows and *Steps to an Ecology of Mind* by Gregory Bateson. Donella Meadows was a pioneering American environmental scientist, teacher, and writer. She was co-author along with her husband, Dennis Meadows, on *Limits to Growth*, first published in 1972. They also co-founded an international network for leading researchers on resource use, environmental conservation, systems modeling, and sustainability that meets every year to advance critical global thinking on issues. *Thinking in Systems* is a great book that describes how systems work in a way that nonscientists can easily understand yet complex and rich in detail to engage scientists as well. The subject matter has been simplified so the average reader can better comprehend it. Gregory Bateson was also another proponent of systems thinking and the intersection of cross-disciplinary thinking. He saw systems as a more holistic means of planning our cities as the way to address and problem solve global environmental issues and other complex planning issues that cities of today are facing. This book is rather dense for the average reader but fascinating in its diversity of ideas and thought proposals.

Ecoliteracy and Systems Thinking

Ecoliteracy, also referred to as ecological literacy, is the ability to understand the role that natural systems play in order to sustain life on the planet and that humans are part of

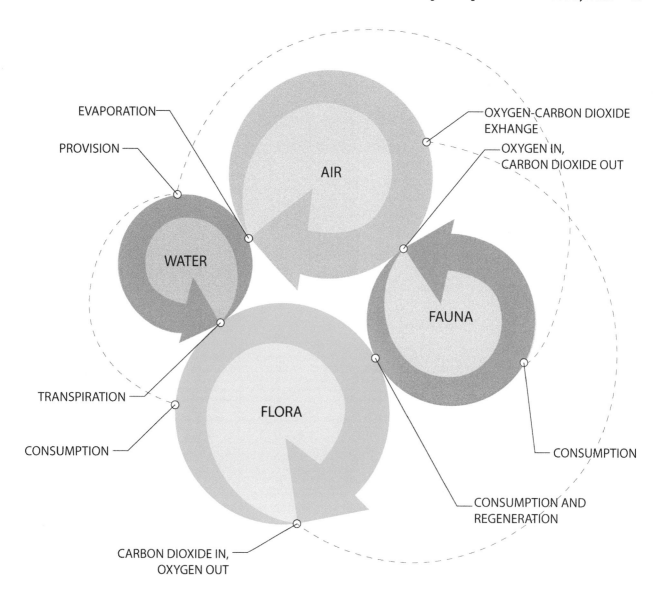

Figure 2.11 Systems thinking in a natural system is interconnected.

Systems Thinking in a Nonnatural System

Parallel to this, in organizations, systems typically consist of people, structures, and processes that work together to make the organization healthy or unhealthy. In a nonnatural system, the parts can be designed for independence or dependence (Figure 2.12). All systems have structure, behavior, and interconnectivity. Non-natural systems can be less efficient than natural systems, as they typically have some measure of "waste" that natural systems do not have. Natural systems may not have an apparent objective, but their outputs may be interpreted as purposes. Human-made systems are made with purposes that are achieved by the delivery of outputs. The key to all systems thinking is the ability for people to be able to consider the potential consequences of their decisions on other parts of the system.

Systems Thinking

Systems thinking is the process of understanding how things relate to one another within a whole system and how each system relates to another. It is based on the precept that the whole is greater than the sum of its parts and that the components of the system are interdependent on each other rather than dependent. There is an interaction and a connection between all of the elements of a system. Systems thinking is a problem-solving approach that views problems as parts of the overall system. In this way, rather than reacting to a specific part, outcomes or events may be determined by a series of connections and developments of a number of parts. It also can be explained that it is not one thing but a series of practices within a framework that views the components of a system are best understood in the context of the relationships they have with each other and with other systems rather than in isolation. In this manner, systems thinking primarily focuses on a cyclical rather than a linear cause and effect. This emphasizes the linkages and interactions between the elements within the system. A systems thinking approach incorporates interdependence of objects and their attributes as part of the holistic system and is considered a dynamic and complex whole (Ackoff 2010 and Checkland 1993).

Systems thinking means that we must think in terms of relationships, patterns, and context. If a system is a set of inter-related elements that make a unified whole, and individual elements such as people, economics, and plants are themselves a system, at the same time these elements cannot be fully understood apart from the larger systems in which they exist. A systems thinking approach helps us to understand the complexity of the world and therefore encourages one to think in terms of relationships, connectedness, and context.

Systems Thinking in a Natural System

Looking at nature through a systems thinking lens would include ecosystems in which various elements such as water, air, and plants work together to thrive or degenerate. An event that would affect one part of the ecosystem would affect the entire ecosystem in either minor or major ways. Sustained life is the property of a thriving ecosystem. Sustainability is achieved through a web of relationships. Thus systems thinking means thinking like an ecosystem in terms of the relationships, patterns, and context when designing the food system for sustainable cities and communities (Figure 2.11).

Nature's patterns and processes include: networks where all living things are interconnected through networks of relationships, nested systems where nature is nested within other systems, cycles that allow for exchange of resources in a cyclical manner and intersect with larger regional and global cycles, flows of energy and matter to sustain life, development and adaptation over time, and dynamic balance that provides resiliency in the face of ecosystem change.

A natural system is organized in a sustainable way that includes interdependence, network of relationships, feedback loops, cyclical flows of energy and matter, recycling, cooperation, partnership, flexibility, and diversity. All living systems develop and evolve; understanding them requires a shift in focus. Within systems, certain relationship patterns emerge again and again in patterns such as cycles and feedback loops. Some aspects of systems cannot be measured but they can be mapped, which can lead to a more comprehensive form of assessment.

political battles in the media and at city council meetings, from angry neighbors surrounding the development, or protests from different groups that are nervous about what damage the change might create.

Another important element to this type of community engagement is the social fabric that is built through trust, understanding differing viewpoints, creating ideas together to implement into the design, and even the simple fact of being heard when something will affect your community. This social fabric is allowing the dialogue to move from potential confrontation to collaborative idea creation and problem solving. The surrounding community was able to give four hours' worth of ideas to the design team through the World Café style meetings on how the community center could be a bridge to the surrounding community. They were able to discuss not just a brainstorm of services that could be provided, but actually what they would like to do there, what value the center could provide to their quality of life, what individuals would be willing to do, what needs the community has that could be satisfied with programs that could take place in the space, and even what meetings could be held there that would bring interaction and social connectedness into the eco village on a regular basis. The new development can now work with folks that might have had worries about the development to build new opportunities that benefit the community as a whole, not just the residents of the eco village.

Instead of being a gated community, it is a porous community that will add richness to the social connectedness of the community. The designers and Habitat for Humanity staff have been able to collect the names of people with their ideas and concerns, then through the trusted relationships that they have, work with the specific individuals in the community to develop projects that help the village add value to the residents and surrounding community. They also know who else in the neighborhood was interested in those ideas and can create teams to make a project happen. This helps ensure the long-term sustainability of the projects that will add value to the community. The people living in the community will have ownership of the projects that they are passionate about creating with others. They are not dependent on the government, the developer, or just the internal residence to execute the services or programs.

Highlights of the neighborhood include:

- 18 homes total, featuring a mix of two-, three-, and four-bedroom units, twin and single-family detached homes with a welcoming, craftsman bungalow feel.
- Super-insulated, high-performance buildings are designed to achieve carbon negative, net-zero energy use and following "Passive Haus" principles and efficient window placement for maximum solar gain.
- Durable concrete and structural panel-built homes are designed to protect residents from severe weather and feature residential fire sprinkler systems, slashing insurance premiums, and saving lives.
- Large solar photovoltaic and solar hot water arrays on each roof, plus a "solar farm" of ground mounted panels on the South hill of the site, offset energy use and utility bills.
- Pedestrian paths connect the entire neighborhood with existing neighborhood, parks and community, promoting zero-emissions transport and the health benefits of walking and biking.
- All paths and drives are designed with pervious paving systems, which along with bioswales and rain gardens prevent runoff and support habitat.
- Rain barrels and cisterns harvest and store rainwater from roofs for the irrigation of shared community gardens and "edible landscapes," providing a constantly renewable source quality local food.
- Affordable for households making 25 to 60 percent of area median income, Habitat homeowners purchase their homes with an investment of up 500 hours of volunteer "sweat equity" and a 0 percent interest, 30-year mortgage held by Habitat. Mortgage payments are "paid forward" to help Habitat build more homes for more people in need.
- Private food landscape zones and public food landscape zones create a community food shed.

Design Team:

Client:	St. Croix Valley Habitat for Humanity
Architect:	Frisbie Architects, Inc.
Landscape Architect:	Gill Design, Inc.
Consultant:	Auth Consulting

EcoVillage Landscape Palette

"Maple Grove"

Sugar Maple - Sienna Glen®
Acer x fremanii 'Sienna'
Excellent branching habit and
very hardy and fast growing.
H 50–60' W 40' Foliage Green
Fall Color Red

Sugar Maple - Fall Fiesta®
Acer saccharum 'Balista'
H 50–60'W 40 50' Green
Foliage, Fall Color:
Yellow/Orange

Berries & Currants
Landsdcape Shrubs

Currants
Ribes rubrum
'Red Lake'

Blueberries
Vaccinium 'Northblue'
Vaccinium 'Chippewa'

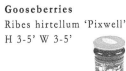

Gooseberries
Ribes hirtellum 'Pixwell'
H 3-5' W 3-5'

Raspberries
Rubus 'Boyne'
Rubus 'Caroline'
Rubus 'Nova'

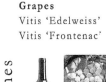

Grapes
Vitis 'Edelweiss'
Vitis 'Frontenac'

Vines

Figure 2.10 The eco village landscape palette integrates edible plants throughout the development that are accessible to the community. Besides the two plant groups shown above, other palette components include edible groundcovers such as strawberries and lingonberries; orchards such as apples and hazelnuts; and perennials in the raingardens for increasing beneficial habitat.

to choose among different landscaping components, including edible components like fruit and nut trees (Figure 2.10). Tenants will also have access to shared community-garden style plots hosted in the common spaces. A shared commercial kitchen will serve both as a site for processing and cooking of produce and as a gathering space for community activities.

There are a growing number of successful eco villages, but the concept is still new to most communities. Any new development in a community can face resistance from people concerned about the change and what it might do to the community. If you add additional change into the situation by developing an eco village that is different than what people are familiar with, the resistance can be even greater. Village Homes in Davis, California, is one of the oldest successful examples of this type of sustainable development model. Because the design is intentionally focused on quality of life design, the developments have been highly successful. Village home property values continue to rise compared to developments around it and there is a waiting list of people wanting to purchase homes there.

The Habitat for Humanity in River Falls, located in the St Croix Valley, has anticipated and understood with empathy this resistance to change. The organization has adopted an integrated community engagement approach for building understanding *and* collecting great ideas from the community. The approach used was of building relationships with the existing community and facilitating conversations that build trust and identify ways to integrate the new community into the surrounding social fabric. The facilitation methods of community conversations using The World Café, Open Space, plenty of coffee meetings in the community, an open door policy at their offices, and weaving the community knowledge into conventional design charrettes has led to a design that not only provides ecosystem services from the land and buildings but also builds long-term relationships in the community. Those relationships are the basis for the development process being efficient by losing the need for big

Permaculture: Is a tool used to design sustainable communities and agricultural systems that are modeled on relationships found in natural ecologies based on ecological and biological principles. It was developed in the 1970s by Bill Mollison, David Holmgren, and their associates as a series of publications in Australia. Current thought is there are now two strands of permaculture, *original* and *design* permaculture. Original permaculture is based on replicating nature by developing ecosystems that resemble their wild counterparts, which is not necessarily a relevant approach for city planning adaptation. Design permaculture looks at the working connections in an ecosystem as the basis for the designed systems and potentially a promising approach for integrating urban agriculture as a system wide green infrastructure component in city planning (Holmgren 2012, Mollison 1988).

For purposes of this book, urban agriculture, edible urbanism, food landscapes, and urban agriculture landscapes will be used interchangeably for policy purposes and project scale discussions, and urban agriculture will be used as the umbrella terminology for any sustainable landscape with an edible component.

River Falls Eco Village, River Falls, Wisconsin

For the St. Croix Valley's Habitat for Humanity, building green means building homes that are environmentally responsible and affordable now and well into the future. River Falls Eco Village is the demonstration project of this green ethos, the first neighborhood-scale project of its kind to combine cutting edge, energy-efficient technology, renewable energy generation, and innovative community design practices into one holistic package (Figure 2.9). Urban agriculture is one of the components of the environmentally sensitive approach and has both private food landscapes and public food landscapes for the entire community. For their own yards, residents will be able

Figure 2.9 The River Falls master plan and a conceptual plan of the various edible landscaping opportunities.

security, food justice, community empowerment, sustainable education, and green job training are just some of the issues they may take part in.

Landscape urbanism: On the other hand, the term *landscape urbanism* has gained momentum in the design industry to describe city building from a more ecological approach to create environmental and social sustainability within a city's network so urban agriculture does fit within that definition but is not necessarily inclusive. Landscape urbanism is a broader way to look at urban sustainability that balances the creation of a resilient and regenerative city landscape. Landscape at this scale encompasses the entire city.

Agricultural urbanism: This has been described as the next big movement for new urbanism by a group of thought leaders and practitioners in the southwest of Canada. The handbook outlines strategies to create agriculture and food precincts and community places where food is celebrated year round. It is rooted in a sustainable food system approach and looks at the issues through an urban design lens that is based on architecture systems.

Bioregionalism: This is the belief that social organization and environmental policies should be based on the bioregion rather than a region determined by political or economic boundaries. A bioregion is an area that consists of a natural ecological community with characteristic flora, fauna, and environmental conditions such as a watershed and bounded by natural rather than artificial boundaries.

Agro-ecology: This is the application of ecological principles to the design and management of sustainable agro-ecosystems. Agro can mean either field, soil, or crop production, so combining it with ecology principles creates a whole systems approach to agriculture and food systems development. This approach is based on the blending of traditional knowledge, alternative agriculture, and local food system experience. It links ecology, economics, and societies to sustain agricultural production, healthy environments, and viable food and farming communities.

Eco village: Eco villages are communities designed to integrate the surrounding ecology into the quality of life of the residents of that development, or village. Eco villages are intentional communities with the goal of becoming more socially, economically, and ecologically sustainable. Eco villages provide the ecosystems services that lead to a high quality of life for the residents. They are designed to save energy, capture and use natural resources sustainably, control the air and water quality, build habitat for a stable ecology on the land, give space for food production, and also provide flexible space for a variety of community interactions.

Garden city movement: A method of urban planning initiated in 1898 by Sir Ebenezer Howard in the United Kingdom. The garden city idea was also influential in the United States for a brief period. Garden cities were intended to be planned, self-contained communities surrounded by greenbelt parks containing a proportionate area of residences, industry, and agriculture. The concept was to produce economically independent cities with short commute times and the preservation of natural open space in the surrounding countryside. This philosophy also offers some insight into developing more sustainable communities that incorporate urban agriculture into the city infrastructure.

Urban Agriculture Landscapes

Urban agriculture landscapes are any landscapes that promote the integration of people, their living environments, and food. These productive landscapes are aiding the promotion of a sustainable urban food system approach. They include food landscapes growing in an urban or peri-urban environment, whether on an income-generating scale or not. An urban agriculture landscape may include food-producing plants and animals or both. It may or may not be based on sustainable methodologies or procedures unless that is part of the landscape's mission and goals, but more and more urban agriculture landscapes are organic or practice ecological best practices that remediate and heal the urban environments they are located in.

When urban agriculture is thought of as part of the green infrastructure of the city, promoting ecological biodiversity and social sustainability, then it is elevated to a sustainable system no matter what the scale, type, or location of the landscape. Examples of urban ag landscapes are diverse and range in size and complexity. An urban agriculture landscape could be a city street planted as an orchard with fruit trees that can be harvested by the local neighborhood or as small as a window box planter in front of a restaurant used by a chef in the daily menu. It could be a 10,000-square-foot community garden on a city lot or it could be a 3,000-square-foot edible garden terrace in an apartment high-rise that includes herbs and vegetables for tenants to grow, harvest, and eat. It could be a one-acre rooftop farm that grows, harvests, sells its produce, educates the community, and trains local youths for green jobs. Or it could be an edible school garden in planters on an asphalt lot that teaches students about nutrition and real food or a vertical garden grown on a blank building façade that has herbs and seasonal fragrance that pedestrians just enjoy as they walk by. It could be a corner plaza with perennial herbs and medicinal plants that citizens and passersby could pick for their own use. It could be a victory garden at the city hall grown for any citizen volunteer to tend and or harvest. The "it could be" list is really as unlimited as there are places and people in the world to think of them.

A few key urban agriculture definitions:

Food landscapes, edible landscapes, productive landscapes, and urban agriculture landscapes: These terms as used in this book are synonymous and are used interchangeably to describe landscapes that promote the integration of people, their living environments, and food. These landscapes are based on an integrated sustainable urban food system approach that fosters the health of the ecology, community, and economic vitality of a city with an ultimate goal of sustainable resiliency. This approach is rooted in environmental ecoliteracy, community interaction, and the integration of a citywide sustainable infrastructure system network. Landscapes can encompass the scale of a city as a landscape or the smaller scale of specific landscapes within the city.

Edible urbanism: The term *edible urbanism* is not an officially coined word for urban agriculture. It could be used to describe planning methodologies for incorporating food sheds and their associated system connections into the infrastructure systems of a city.

Agtivist: People who champion urban agriculture in their cities and towns, many of whom are focused on advocacy, changing policies, grassroots action, and the locavore food movements. Access to real, healthy food, farmers markets, food

the natural world. If we are to build and plan for more sustainable communities we must design them in a way that they will not destroy the natural environment on which they depend. Ecoliteracy is the understanding of the connection between ecological health and human health. There is a huge gap today in most communities and schools about the role of real food and its connection to health.

One nonprofit organization that saw the need for this type of environmental thinking education is The Center for Ecoliteracy located in Berkeley, California. Cofounded by philanthropist Peter Buckley, physicist Fritjof Capra, and think-tank director Zenobia Barlow the organization was founded to provide the means for teaching ecological principles and system thinking for K–12 education. In the book *Ecoliterate: How Educators are Cultivating Emotional, Social and Ecological Intelligence* (Jossey-Bass 2012) the Center highlights the 5 vital ecoliterate practices as being:

1. Develop empathy for all forms of life.
2. Embracing sustainability as a community practice.
3. Making the invisible visible.
4. Anticipating unintended consequences.
5. Understanding how nature sustains life.

The Center's projects include initiatives such as the Rethinking School Lunch program that was created as collaboration with the Chez Panisse Foundation and the Berkeley Unified School District to provide local, seasonal, and sustainable meals for children combined with experiential learning in the school gardens, kitchen classroom, and cafeterias. Using a systems approach Rethinking School Lunch program provides a planning strategy for a more sustainable lunch program in schools. This initiative has caught on around the country. Amy Kalafa's book *Lunch Wars* offers ideas on starting a school food revolution in your own community to address children's health and is an excellent example of how this idea for rethinking ecology and food is starting to gain traction. In Chef Jamie Oliver's show *Food Revolution,* Jamie has made his own quest for tackling this subject and documenting the struggles as well as the health issues children and communities are facing. Jamie has a website and blog that provides a network for communities to learn from and to share their experiences as a community.

The most recent initiative by the Center for Ecoliteracy is "Smart by Nature: Schooling for Sustainability," which identifies four potential pathways to integrate sustainable learning in schools. Its publication *Big Ideas,* written for K–12 educators, offers guidance in graphic and sophisticated simplicity to utilize as a conceptual framework for an integrated curriculum. Programs like these reinforce that education begins early and that school gardens offer a lifelong foundation for learning about food, health, and sustainability in a new integrated model based on a systems thinking approach. This education, in turn, provides the foundation and building blocks upon which our communities and cities grow. Urban agriculture and food landscapes are ripe to play an important part in the education of the connection between food, health and ecology to society as a whole.

Interconnectivity + Interdependence—How Systems Work

Interconnectivity and interdependence are two of the key ways systems work together or within themselves. Interconnectivity is the ability to connect reciprocally. Interconnectivity deals with systems in dynamic equilibrium. Used in numerous fields such as cybernetics, biology, ecology, network theory, and nonlinear dynamics, it is a concept that can be summarized as follows: all parts of a system interact with and rely on one another simply by the fact that they are components of the same system. It also understands that a system is difficult and even impossible to analyze through its individual parts considered alone since a system is more than the sum of its parts.

Interdependence is the dynamic of being mutually and physically responsible to others while or through sharing a common set of principles with others. It differs from dependence because dependence implies that each part of a relationship cannot function or service without or apart from the other one. An interdependent relationship is described as one where all participants are emotionally, ecologically, economically, and/or morally self-reliant, while at the same time they are responsible to each other. This type of system relationship can also be described as an entity that depends on two or more cooperative and autonomous participants such as a coop. Relationships of interdependence seek to operate by recognizing the mutual "value" in each entity and weave them together in a synergistic manner (Kelly 1994). A system with interdependence and interconnectivity equals a system with dynamic cooperation. Connecting systems through the interdependent relationships between them will allow for a flexible, dynamic, and cyclic-oriented network of systems that function on a more integrative level. Because of the complexity of the information, multilevel systems and new graphic models to convey complex scientific data need to be created. Currently, these do not readily exist as reference guides beyond a few two-dimensional models because system criteria vary physically, politically, economically, and culturally from city to city. City and regional planners should collaborate with a network of academic disciplines to create dynamic system relationship testing models that would assist cities in incorporating environmental sustainability, food security and food justice, and community health into sound urban planning models.

Stephen M. R. Covey in *The Speed of Trust* (2006) tackles the issue from the angle of how organizations and businesses can thrive with productivity and satisfaction through a relationship built on trust. He makes a case that this relationship is the basis for the new global economy so that layers of laborious bureaucratic check and balance processes can be eliminated.

A recent documentary called *Connected* by Tiffany Shlain is an inspiring dialogue about how we connect as humanity with each other and how technology is changing us. The film offers starting points for further dialogue through its interactive website (http://connectedthefilm.com).

PART 2
CITY PLANNING STRATEGIES FOR URBAN AG LANDSCAPES

As we plan for more sustainable cities in the twenty-first century, urban agriculture's role as an important and viable sustainable food system must begin to be included in city planning processes just as transportation, water, and energy systems currently are. This will require developing a new relationship between city planning and urban agriculture. This relationship should focus on the evolution of an urban food system approach as the foundation to invite food back into our lives, our homes, and our cities.

> The development of the food system as an integral part of the city begins by asking this question: How do we create conditions that will allow for each community to feed itself? This will require thinking of food as part of an ecological model.

Harnessing an Urban Food Systems Approach through Urban Agriculture

Many cities in the United States do not have legislation that protects agricultural land around cities; nor, if they do allow urban agriculture as an allowed land use on a small scale, do most allow it on a larger scale than one acre. Grassroots change is being led by cities such as San Francisco, New York, Chicago, Seattle, Detroit, and Baltimore. As the grassroots awareness for developing a new relationship between agriculture and city planning evolves, the integration of a more sustainable food system philosophy, principles, and practices, rests solely on new development and grassroots activism to facilitate the dialogue. This dialogue between design and planning professionals, local governments, and the community must focus on the need for integrated infrastructure, urban open space design, human scale agriculture, and the creation of more flexible policies that allow for food landscapes within a city.

60 Designing Urban Agriculture

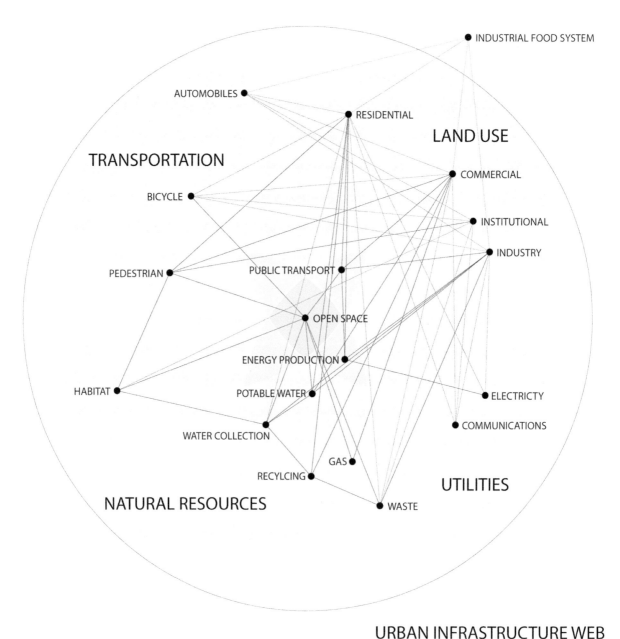

Figure 2.13 A systems graphic shows the integration of city systems for a greater whole, resulting in an urban infrastructure web.

This new urban planning approach must consider the following:

1. Integrated systems thinking must focus on solutions based on the interconnectedness of the systems as a whole unit rather than as independent units (Figure 2.13). The strength and efficiency of the connections increase the more integrated the node is within the web. The network of the systems is a vital component that functions better with integration and interdependence than with separation. Note the industrial food system node is an outlier with nominal connections to the urban infrastructure system.

2. The integration of natural and urban systems into city infrastructure needs to include urban agriculture models, also referred to as urban food systems (Figure 2.14). Urban agriculture concepts, especially nonconventional ones, offer new ways to rethink city infrastructure systems to move water, energy, people, and materials

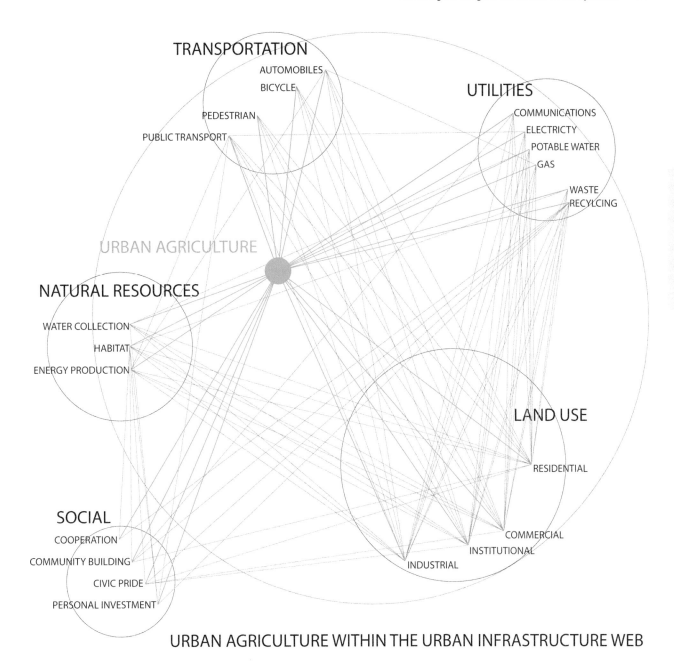

URBAN AGRICULTURE WITHIN THE URBAN INFRASTRUCTURE WEB

Figure 2.14 A systems graphic integrating urban agriculture system into the city infrastructure.

that will benefit both the city and the regional systems. Infrastructure systems become more efficient as integration increases. The inclusion of urban agriculture creates an additional node within the infrastructure web, strengthening the system as a whole and reducing the need for outside inputs.

Current planning and city development departments such as transportation, public works, public utilities, urban forestry, and building departments function as either independent or co-dependent compartments, with each department and subdepartment functioning as their own fiefdoms. This current relationship does not foster cooperation or collaboration and, in fact, inhibits both by creating layers of bureaucracy and red tape that preclude an integrated systems thinking approach.

Creating a Policy Framework for Urban Agriculture

When it comes to the policies that promote urban agriculture and embrace food in the city as a sustainable-based approach that benefits the city and community, general planning issues must be addressed and an overall food policy framework must be integrated into a city's planning process.

The policy framework in the urban planning process includes the following:

- ***Establishing a regulatory and legislative framework in every city that supports urban food and agriculture.*** This affects the ability to streamline permits to understanding that zoning regulations currently do not provide for a more integrative systems approach. Some cities, such as San Francisco, New York, Detroit, and Baltimore, are on the front lines of these issues and are making changes at a rapid pace to allow for these new landscapes to continue to set roots in the urban realm. This is currently being addressed from both a citywide interdepartmental approach as well as from a grassroots organizational approach that clearly highlights the issue to the community affected by the city's decisions. "Get involved" and "Understand the issues" are the two messages that clearly are important for both sides.

- ***Setting food and agriculture goals early in the planning process with city, stakeholders, and communities.*** It is important that all stakeholders are included in the beginning of the process. This builds consensus early on and conserves energy that would likely be expended if not done upfront. Communication is key, and various tools such as World Cafés, community forums, community workshops, and others can be employed.

- ***Including local governments in the process.*** Local governments are on the frontlines of sustainable community planning in that land use is paramount to any strategy in addressing climate change, affordable housing, economic development, and food and agriculture, among others.

- ***Establishing a research component through government, nonprofit, or for-profit enterprises that is available for cities to use as a toolkit.*** We need more dialogue, research, and strategic thinking on all levels of information sharing for both citywide departments and for food entrepreneurs—both large and small. Most urban ag landscapes have sprung from a grassroots level, community activism, or innovative thinkers who are one step ahead of everyone else. There is still a need to develop local and regional connectivity—a scale aggregate foundation is desirable to increase urban agricultural landscapes throughout the city, not just in a few areas. One example on a national scale with a tremendous amount of free legal information is The National Policy & Legal Analysis Network to prevent childhood obesity, or simply NPLAN, whose goal is to connect research to action. Information is available to assist legislators, policy makers, teachers and school districts, parents and caregivers, public health professionals, and researchers and academics. Categories include foods, schools, kids, and communities. An example of the type of information available is the templates for community garden business plans that can be downloaded and then molded to the community garden's specific vision, goals, and objectives. The Center for Ecoliteracy described earlier in this chapter also offers free publications and materials for K–12 educators.

The policy development process needs to be integrated early on as a key element of creating a local government policy framework that promotes urban agriculture. The following activities are important to include in this part of the process:

- ***Support for a community visioning process.*** Without key community and city stakeholder involvement and engagement in the urban agriculture vision, creating policy that promotes urban agriculture will ultimately fail.

- ***Setting goals and objectives within the master planning process.*** Setting the goals and objectives early on in the process provides for a more holistic and multidimensional system that brings up issues of connectivity with other, more traditional city systems typically thought separate from urban agriculture. For example, how does distribution of produce connect reciprocally to the transportation systems within a community? Or, how does incorporating a zero-waste program on an urban ag project connect reciprocally to the neighborhood's waste-removal system? The master planning process can help to establish the urban ag system as an integral part of the overall city plan.

- ***Creating awareness and momentum through the local area planning process.*** Without awareness there will be no momentum in building a successful urban agriculture vision. Creating awareness though the planning process includes creating documents on recommended urban agriculture guidelines, holding educational forums, providing for city council informational meetings, and other ways that provide a platform to inspire others to build on the vision. This type of visioning process builds consensus, increases stakeholder support, and increases the system complexity to better shape and realize the vision.

- ***Creating awareness through servicing and infrastructure planning.*** Without awareness of how urban agriculture is a vital part of the city's green infrastructure systems, urban ag will remain relegated as a temporary activity and not a permanent and vital system activity for the communities they serve. Creating awareness of the metrics with all departments and city leaders is an important part of awareness. Private planning organizations within a city such as San Francisco's Planning and Urban Research, or SPUR, can help to document activities and provide impetus for initiatives that can push the metrics and pilot projects forward into the public realm.

- ***Including provisions within the city's planning criteria for addressing climate change.*** Most cities have begun to address climate change through their own planning criteria. Urban agriculture needs to be included into these documents and guidelines as part of the environmental solutions that address ecological degradation, pollution in the atmosphere and ground, heat island effects on temperature, stormwater treatment and management for watershed vitality, human health as it relates to disease, food security and food justice, and lastly, how these environmental solutions include social solutions that address community vitality and prosperity.

- ***Creating awareness through the standard design development process in implementing the city's broad policy framework.*** As a project moves forward from planning to permits, there are many opportunities to begin to address urban agriculture as a project component through these steps such as planning reviews and entitlements, and through sustainable rating system commitments such as LEED, Green Point, and SITES to name a few, and local ordinances. In Chapter 3 we will take a more in-depth look at the design framework process for urban ag landscapes. Chapter 6 will take a closer look at specific policies and policy development required for promoting sustainable urban agriculture and will provide a glimpse of cities that are changing the face of urban ag policy around the country.

15 Principles for a Systems Thinking Approach for Urban Agriculture

To build an integrated systems thinking model for urban ag landscapes, let's begin by first establishing 15 principles that promote a more integrated approach.

15 Urban Agriculture System Principles
1. Promote biodiversity.
2. Increase food security and food safety.
3. Incorporate education and outreach for awareness.
4. Be climate adaptive for environmental resilience.
5. Maximize water accessibility, availability, and quality, and address the value of one drop.
6. Maximize waste and energy effectiveness.
7. Provide for soil resiliency and soil health.
8. Develop a systems network that is both regionally and locally appropriate.
9. Promote social responsibility.
10. Protect and increase human health benefits.
11. Provide for the connection of people with nature to the enrichment of both.
12. Foster community, placemaking, and social resilience.
13. Develop dynamic connectivity between the human and ecological systems.
14. Promote sustainable economic benefits and opportunities.
15. Increase the treatment of waste as a resource for a zero-waste outcome.

Urban Agriculture Methodologies and Tools

Where food has been included in city planning objectives in the past, it has focused primarily on the development and creation of community gardens and spaces for farmers' markets. However, there are so many more opportunities to capitalize on that food, and its social and economic dimensions, can offer a community. This section describes methodologies and tools for planning urban agricultural landscapes.

Placemaking

Placemaking is both an idea and a tool that considers the context of creating communities that are desirable places to live. It is based on the idea that places are complex and dynamic systems. As a tool, placemaking is based on listening to, looking at, and asking the questions that matter of the community and users when designing or planning spaces.

There are a variety of placemaking communication tools available for community engagement that can be adapted to include urban agriculture as a systems component of the placemaking dialogue. One example is the *Placemaking Tools for Community Action* guide that is based on four system layers identified as social, economic, built, and natural. It takes the reader through a series of steps that helps to identify the specific goals, determine the applicable planning process, create a community value system, and explore specific planning tools available to utilize for the project. Some of these tools are asset mapping, visualization, or predictive modeling including a rundown of available and emerging planning tools on the market to consider. (Boyd, 2002)

- ***Shaping a community's identity and activity patterns.*** Urban agriculture has the potential to shape the physical framework of a community's daily life. When scale is allowed to be multifaceted, a systemwide interdependent approach is intertwined into the existing framework, and zoning is allowed to be flexible enough to foster new system connections not previously envisioned. Urban ag can shape streetscapes, neighborhood gathering and circulation, create cultural landscape identity, and provide multigenerational aspects not previously planned for or available to a community.

- ***Identifying social and well-being metrics related to food.*** Metrics are needed that acknowledge the many aspects of food and agriculture that are central to fostering health and happiness in our lives. Food is central to human biology, sociology, and psychology. Emerging tools such as the Sustainable Sites Initiative, or SITES rating system, now in its pilot phase, are beginning to measure environmental and social metrics as they relate to designed environments and identify their value to human health and well-being and ecosystem services. Urban agriculture needs to be incorporated into the metrics data of this tool and other emerging tools.

The SITES rating tool was developed by the American Society of Landscape Architects, the Lady Bird Johnson Wildlife Center, and the US Botanic Garden, along with a number of public and private sector partnerships to develop a sustainability tool that looks at landscape at a regional and site specific scale to evaluate and increase environmental and social sustainability. Other sustainable development resources to look into include One Planet Living and Biomimicry.

Food as a Platform

Food can become a platform or layer from which we address other important elements of community, ecology, and livability, including the physical, social, economic, cultural, and environmental health of the city. As we discussed earlier, everyone needs food to survive, enjoys food as nourishment and celebration, and is connected to others by food as we celebrate or gather for a meal. Food is the gateway to the stakeholder conversations between city, community, and project developer.

Integrated Urban Design and Physical Planning

Integration of urban agriculture into the physical planning and design of communities requires attention to both the urban design process and its physical outcomes. Where possible, combine urban agriculture components with complimentary uses. Incorporate multi-compatible activities such as recreation opportunities, both passive and active, and look for maximization of the experience through education, training, retail, and viewing opportunities. Urban ag landscapes should also be designed to maximize the opportunity for connecting people to nature.

Furthering the idea of urban ag as part of the urban design and physical planning process, on a large scale, the urban ag system, aka the food system, should be considered as part of the green infrastructure of a city planning system's matrix. Urban agriculture has the potential for integrated connectivity with a city's infrastructure systems—such as stormwater, waste, energy, open space, and natural resource management.

There are four key principles for integrated infrastructure systems:

1. Interconnect the systems so that the output of one system is the input for another system and treat all waste materials from the systems as resources.
2. Generate value and revenue from these collected resources.
3. Achieve multiple benefits from each system where possible.
4. Position the resource producers and users near enough to each other to facilitate resource exchange (de la Salle 2010).

Ecosystem Planning

The role of ecology is of utmost importance when planning for the integration of urban agriculture in urban environments. Without an ecological foundation, urban ag remains conventional and a nonsustainable part of the city landscape. It is important to look for the interconnection between the ecological systems, social systems and the urban ag systems being planned for. Examples of ecosystem planning integration are the planting of crops specifically for wildlife habitat, or in preserving and creating riparian areas to manage stormwater and support wildlife within the urban ag landscape design.

Urban Open Space Design

Some cities already acknowledge community gardens as part of the city's overall open space system and planning policy is generated from their open space guidelines. Seattle's P-Patch community gardens were developed as a key component of Seattle's open space planning policy. Locations and community-oriented aspects vary with each neighborhood garden. Orchards and habitat corridors of beneficial plant communities that add to the sustainable ecosystem function for urban farms and gardens could be introduced as elements of the open space as parks, trails or community recreation nodes and learning centers.

Circulation and Connectivity

The idea of connectivity extends to circulation of people such as trails, walks, and paths that link food landscapes to other community amenities including other food landscapes, activity functions and open spaces such as parks or recreation centers. Vehicular circulation should address ag function support needs such as volunteer or employee accessibility, storage, distribution, and processing aspects, not to mention deliveries to markets or organizations such as food banks or school lunch programs.

Scale Development and Location Maximization

Scale development involves integrating urban agriculture into all scales of urban development from building to street, neighborhood to cities, balcony to front stoop, and beyond. For true integration as a citywide system approach the smallest landscape scale to the largest must be included and planned for. Where possible, combining urban agriculture components with complimentary uses is a key to maximizing scale benefits and connectivity.

Scale Aggregation

Scale aggregation is a land management technique that consolidates numerous smaller farms in urban environments to create a more viable economic return through their collective repositioning as a larger citywide network entity. The larger scale entity allows for a degree of economic return not possible with smaller entities. Big City Farms in Baltimore is an urban ag example of this practice.

Human Scale Agriculture

This scale of food landscape includes urban agriculture at a personal level and scale that enriches the quality of life for each resident of a community. Examples would include backyard or front yard community scale agriculture, or CSAs, community landscapes that enrich a community such as ag crops along a public streetscape, or a victory garden managed by the neighborhood community. Edible estates and urban homesteading are other examples of this scale where a family or homeowner grows crops or raises chickens to supplement their diet with healthy, fresh food. Even window boxes filled with herbs or micro-greens that can be harvested for use in a family meal are human-scale agriculture to be enjoyed and savored singularly or with family and friends.

Artisan Agriculture

Because conventional agriculture does not integrate easily into the urban fabric, the artisan model is being used to describe a newer flexible and adaptive approach. Since artisans were the primary producers of goods before the industrial revolution the term also takes on a more hands-on individual oriented crafted technology and product, instilling a higher, more unique level of quality in contrast to the mass production of conventional agriculture. Many of these items are being sold at farmers markets and gourmet stores or even used in specific restaurants with the local farm produce used noted on the daily menu.

Cowgirl Creamery's cheese is an example of an organic cheese company located in Pt. Reyes Station a small town in west Marin County, a county that is known for its sustainable farming. Started in 1997 by Sue Conley and Peggy Smith, who built a small plant in an old barn to create handcrafted organic cheese made with organic milk by their neighbor Strauss Family Creamery. Champions for sustainable practices and environmental stewardship, they also support their cheese-making friends in being sustainable land stewards. A sustainable success story, they continue to make their own small collection of cheeses totaling about 3,000 pounds per week, but they also distribute amazing artisan cheeses from their cheese-making friends from America and Europe.

Identifying and Connecting to Productive Distribution Potential

The incorporation of urban agriculture into the city framework includes looking at the distribution of the urban agriculture products and resources. These distribution outlets include but are not limited to retail sources such as specialty stores, restaurants, community supported agriculture programs, food banks, farmers markets, community festivals, special events, charity organizations that feed the homeless, and school lunch programs.

Connecting Urban, Peri-urban, and Rural to a Regional System

Connectivity of the various natural and built resource systems provides for a more integrated resource management structure that current conventional agriculture and current city planning systems do not consider. The benefits and value of a regional system add to the ecological, social, and economic sustainability of the city and the outer city boundaries.

Permaculture

Modeled on relationships found in natural ecologies based on ecological and biological principles one design framework for creating integrated and sustainable food ecosystems is Permaculture Design. As defined by Bill Mollison in *Permaculture: A Designers Manual,* "Permaculture is a design system for creating sustainable human environments" (Mollison 1988, preface page ix). Permaculture provides a foundational ethic coupled with a set of principles that guide any designer toward creating a landscape that emulates a healthy ecosystem. As designs are put to paper, the three ethics and 12 design principles can act as a sort of proof to check how well your formula works in creating a sustainable and integrated system (Holmgren 2012).

At each phase of the design process you can check if it does the following: takes care of the earth, takes care of the people, and shares the abundance with others. If it does, your design is based on a holistic, integrated ecological ethic found in healthy systems. Working from the foundation of these ethics, one can move through the 12 design principles to design any size system to be ecologically sound, highly abundant, diverse, and regenerative of ecosystem services in your community.

The 12 principles as written by David Holmgren, Permaculture Co-founder, are the following:

1. Observe and Interact: Beauty is in the eye of the beholder.
2. Catch and Store Energy: Make hay while the sun shines.
3. Obtain a Yield: You can't work on an empty stomach.
4. Apply Self-regulation and Accept Feedback: The sins of the fathers are visited on the children unto the seventh generation.
5. Use and Value Renewable Resources & Services: Let nature take its course.
6. Produce No Waste: Waste not, want not. A stitch in time saves nine.
7. Design from Patterns to Details: Can't see the forest from the trees.
8. Integrate Rather than Segregate: Many hands make light work.
9. Use Small and Slow Solutions: The bigger they are the harder they fall. Slow and steady wins the race.
10. Use and Value Diversity: Don't put all your eggs in one basket.

11. Use Edges and Value the Marginal: Don't think you are on the right track just because it is a well-beaten path.
12. Creatively Use and Respond to Change: Vision is not seeing things as they are but as they will be.

Of course, this list is just short statements that represent a thought process and analysis tool that help you think through your design without giving the "correct" design data. This allows you to create sustainable patterns in a complex system within the site-specific considerations of your project. They are simple statements that, when applied systemically to a design, can create profound results.

The design principles are a framework for designing human ecological systems of any kind. They don't provide the answers to a specific design situation but a design approach to ensure what is created is ecologically sound and abundant, while improving the natural capital and resources for the community. It is design thinking and systems thinking with ecological wisdom wrapped up together. It allows a framework to improve your site specific designs of space, aesthetics, client needs, and community needs, while ensuring that the patterns in the design that manifest are creating a healthy system—a system where the gardens, the streets, the politics, the community at large, the gardeners, the behaviors, the exchange of goods and services, and many of the other things that make up a food city are working together in symbiotic relationships with each other. Basically, it promotes getting along well and helping each other all have abundance and a high quality of life.

To be an expert at this takes numerous efforts and study of examples of others' work in order to get a sense of the complex layers that work together in these systems. However, anyone can learn the principles and start to work with them to improve designs. Inherent in the design principles is a phased approach of zones and layers of simplicity working together to create diverse designs. It moves from observation through planning and simple gardens to start with, all the way to complete ecosystem regeneration. You can access it at the level needed for your skills and the needs of the project. The permaculture community of designers is full of opportunities to hire consultants, to mentor other experts, and to gain hands-on experience of projects in numerous capacities. It is now relatively easy to find someone in the local community to help think through how the principles could apply to the food system you are planning and designing (Holmgren 2012).

Urban Design Strategies for Urban Agriculture Landscapes

A framework of urban design strategies can be used to aid in creating urban agricultural landscapes that promote ecological biodiversity and social sustainability in urban environments no matter what the scale, type, or location of the landscape. Incorporating a systems thinking approach that considers appropriate methodologies and tools, creativity and innovative technologies, combined with the urban design strategies that follow, urban agricultural landscapes can flourish and aid in the evolution of the sustainable city. Most of us would think of eating and farming and perhaps the buying of fresh food at farmers markets if asked to identify the elements of a food system. For a sustainable urban food system there are actually eight components to understand when diagramming the food system for a project that will help to determine appropriate planning and design strategies to consider when starting the design process:

1. ***Growing and operations:*** the growing, raising, and managing of food landscapes.

    This is primarily an on-site system with input and output connections.

2. ***Processing:*** the process of taking raw food products and refining them into a more complex product. This includes baking, grinding, preserving, and other methods of transforming raw ingredients through human actions.

    This can be an on-site or off-site system with input and output connections.

3. ***Distribution and storage:*** the distribution and storage of the raw and the refined foods products.

    It can be an on-site or off-site system with its associated input and output connections.

4. ***Selling and buying:*** the purchasing, wholesaling, and retailing of food products, which may also include community trading and bartering types of food exchange and programs such as CSAs.

    This may be on-site system as in farm stands or community bartering but it may be an off-site system such as farmers markets, restaurants, or retail stores with the variable associated input and output connections. Many urban ag landscapes do a combination of on-site and off-site systems.

5. ***Eating and celebration:*** the eating and enjoyment of food whether it be for nutrition and sustenance or feasts and celebrations in either public or private places.

    This is typically an off-site endeavor from the food growing operations but on-site consumption is also a consideration of the system inputs and outputs.

6. ***Waste and recycling management:*** the utilization, management, and diversion of the organic waste created by food growing, food processing, and food consumption. In a zero-waste environment, it also includes all waste-stream management throughout the entire food system process and thinks of all waste as a resource.

    This is both an on-site or off-site consideration when addressing system inputs and outputs.

7. ***Education and branding:*** the teaching and outreach of the food landscape and food system ecological, social, and economic benefits to both private and public sectors.

    This part of the system includes on-site education, training and mentoring of students, community members, and staff as well as the outreach associated with the marketing and branding of the urban ag development to address ecoliteracy and benefits provided to community and city. There are both on-site aspects and off-site aspects to the systems for education and branding.

8. ***Policy and advocacy:*** the creation of policies, guidelines, codes or ordinances in support of urban agriculture that provide opportunities for food system (Figure 2.15) landscapes to prosper.

    Sometimes advocacy is necessary for the policies to be put in place before food landscapes can be constructed. Sometimes free expression highlights the need for policy change and new directions. There is a relationship between grassroots-driven needs and community and city cooperation. The input and output connections for policy and advocacy will also need to begin considering how the urban ag connects to other systems in the city since this is not in place in most cities.

Planning Strategies for Urban Food Systems   71

1. Growing and Operations
2. Processing
3. Distribution and Storage
4. Selling and Buying
5. Eating and Celebration
6. Waste and Recycling Management
7. Education and Branding
8. Policy and Advocacy

The urban design strategies into which urban agriculture systems can be integrated include the following.

Figure 2.15 The eight components of a sustainable urban food system.

Buildings of Various Types, Forms, and Function

All building types are able to integrate urban agriculture systems into their system matrix through both form and function (Figure 2.16). This encompasses not just as a food support use but also an experiential component. The building program, facades, roofs, and design character are all part of the strategy. Building typologies from all land use types can be rethought to include food system components and integration with the architecture systems. Urban agriculture can part of the building's roofs and walls such as rooftop farms and living edible walls; or, part of a building's interior program such as a processing operation like coffee roasting, a bakery, a microbrewery, or fruit-preserving operation. Widening the building system to include the landscape property provides even more opportunities such as parking lots and streetscapes designed for orchards, backyard or front yard edible programs, community open space gardens and workplace company gardens, or crop-planted infiltration landscapes used for soil remediation and stormwater management purposes. Any building and site has the opportunity to integrate urban agriculture.

Figure 2.16 Urban agriculture can be integrated into all parts of a building and site such as roofs like the Gary Comer Youth Center.

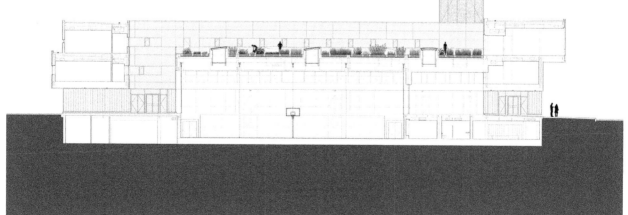

Parks, Plazas, and Open Spaces

All public open space developments such as parks, plazas, recreation centers, vacant lots, and leftover residual open space parcels offer opportunities for community-based urban agriculture, public/private partnerships, festivals that focus on building community through food, and places that can be integrated into the city's open space system (Figure 2.17). As part of the open space system, urban ag landscapes add to the performance of the ecological systems that address sustainable issues for a healthy and vital city infrastructure. The resulting urban agriculture landscapes have the ability to add to a city's food security aspects, increase benefits for human health and well-being, increase the ability for circulation and connectivity, add to the local habitat corridor network, assist with stormwater management, include interaction through public art and play, and provide opportunities for education that addresses ecological, sociological, and economic stewardship to benefit city and community.

Figure 2.17 Lafayette Greens is an example of a park that integrates of urban ag as part of the open space system.

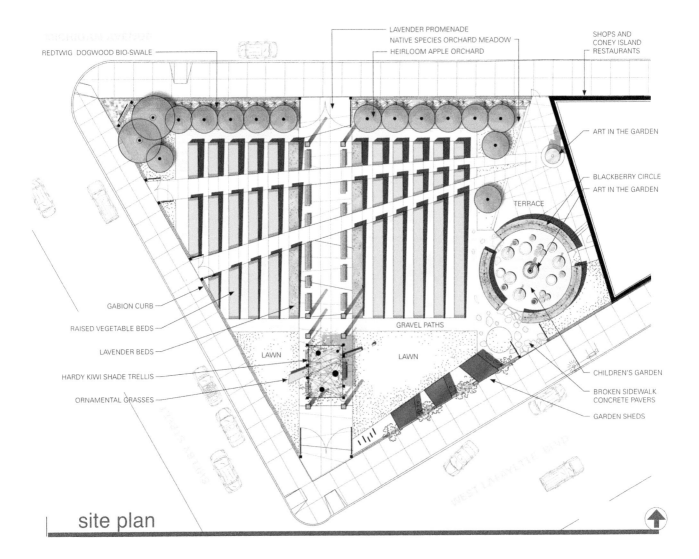

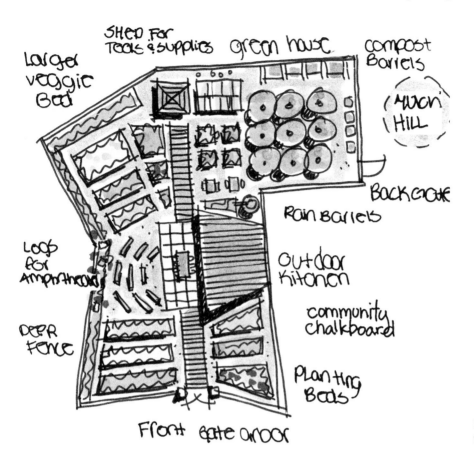

Figure 2.18 A school garden plan from Miller Creek Middle School.

Community, Demonstration, and School Gardens

All community, demonstration, and school urban ag landscapes should be considered part of the city's open space system. Connectivity to other community amenities should be maximized, along with education, stewardship, social needs, human health, and well-being aspects. School gardens (Figure 2.18) with a kitchen component offer not only a productive edible landscape but the potential to enhance curriculum based on age-appropriate earth sciences and nutrition. These garden types can also provide training and mentoring opportunities within the greater community.

Streets and Transportation Infrastructure

Food landscapes integrated into smart sidewalks, trails, paths, medians, and streetscapes (Figure 2.19) offer a green component to the transportation infrastructure of a city and offer the ability to make streets safer, walkable and pedestrian oriented precincts instead of being dominated by the automobile. Streets as part of the city's circulation grid offer the opportunity to rethink the street in a greener light through food system development. As long as the streets can continue to convey the necessary traffic, they can be redesigned to include stormwater management swales, community gardens, recreation and play areas, picnic nodes, public art, habitat corridors, and food. If large enough, they could provide for linear community garden lots, orchards, or even crops for land remediation and screening. Transit corridors and greenways can be designed with ecological and cultural purposes and offer opportunities for multifunctionality and connectivity.

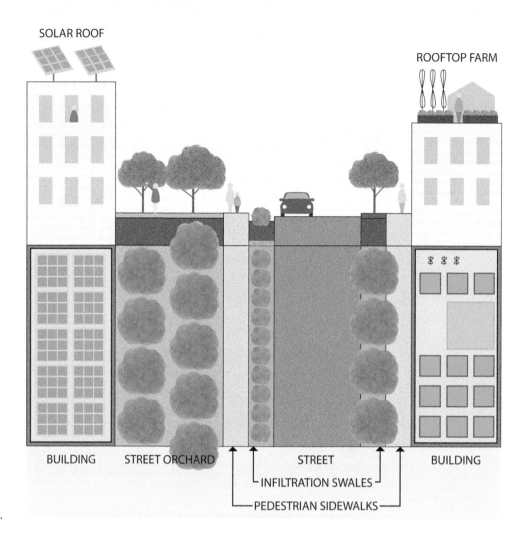

Figure 2.19 Streets and greenways offer places to serve as urban ag landscapes.

Green Infrastructure Systems Including Energy, Water, Waste, Solid Waste, Communications

Through systems thinking applications such as ecosystem planning, circulation and connectivity, and integrated infrastructure (Figure 2.20), urban agriculture has the ability to rethink the city as a sustainable interdependent series of connections and systems that provide for a multifaceted, multiperforming, multiconnected activities and functions as well as for sustainable climate mitigation programs.

Figure 2.20 Green infrastructure opportunities for food landscapes abound in cities and towns.

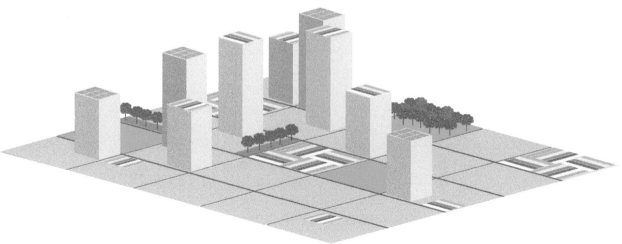

Verge Sidewalk Garden, Charlottesville, South Carolina

One way resourceful urban farmers are making use of under-used land is by planting in the land between sidewalks and streets (Figure 2.21). Typically, only lawn is grown in these spaces, and rarely anything edible. But just because the spaces are small does not mean they can't be productive. One example is the "verge garden" (Figure 2.22) created by horticulturalist Elizabeth Beak and landscape architect William Eubanks outside their Charlottesville, South Carolina, home. Although they encountered some resistance at first, they eventually acquired both a permit and their neighbors' approval. The city is even working toward developing guidelines for urban agriculture as a result.

Figure 2.21 The garden facilitates community and neighborly interactions.

Figure 2.22 The sidewalk garden adds a layer of biodiversity to the residential neighborhood and takes advantage of the front yard's better solar orientation to grow food.

Ecosystem Strategies for System Integration of Food Landscapes

- ***Habitat creation for beneficial insects and pollinators.*** Incorporating a plant palette of native and climate adaptive plants suitable for the local environment encourages the presence of beneficial insects and pollinators necessary for productive edible landscapes. Habitat creation works best when landscapes are connected either as patches, niches, or corridors to provide for wildlife movement and protection. When too scattered or too small of an area and without other areas nearby, productive landscapes suffer from pests, diseases, and lack of pollination.

- ***Habitat corridor connections/network.*** Incorporating linear greenways and transit corridors with urban ag landscapes provides for a larger habitat corridor network that can potentially tie into the overall network of a city's open space system. This adds the potential for better ecological functions and connectivity of the ecosystems resulting in more optimum performance results.

- ***Stormwater harvesting/management for agricultural irrigation.*** Water conservation and stormwater management for water quantity to reduce flooding has become one of the leading resource management issues in a city. With nonpermeable streets, sidewalks, and buildings composing 85 percent of a city's makeup, providing for permeable surfaces and collecting and harvesting rainwater is important to managing this precious natural resource. With the integration of urban ag landscapes into the city fabric, more land is turned over to permeable surfaces, and stormwater from streets and rooftops can be repurposed for food landscape irrigation and water infiltration into the local watersheds.

- ***Stormwater quality.*** Stormwater management for water quality can also be integrated into urban ag landscapes, which have the dual ability to remediate soil and water through infiltration and treatment techniques adding to the ecosystem function and performance of the watershed and top soil nutrient renewal.

- ***Soil health to increase food vitality, quality, and urban erosion.*** The Earth's crust is a six-inch layer of soil that is the breathing lungs of the entire planet. Through past development and conventional framing practices, the health of the soil has become compromised and degraded. Through urban agriculture landscapes integration of ecological based practices such as plant choices, sustainable installation techniques, and organic amendments including compost, the soil can be revitalized adding to food vitality, safer food, and also add to soil's ability to reduce erosion in urban environments. Erosion is a leading factor in increased city flooding and non-vegetated wastelands that do not support living systems.

- ***Increased biodiversity.*** Biodiversity is the existence of a wide variety of plant and animal species in the environment. In ecosystems biodiversity is the source of endurance and resilience. Without biodiversity stability an ecosystem can be jeopardized. By restoring native habitat, removing invasive species, adding plants that attract beneficial habitat and pollinators, the caring for, planning, and monitoring of edible landscapes can add to the increased biodiversity within the city.

- ***Carbon sequestration and air quality.*** The trees and plant biomass of urban ag landscapes add to the city's ability to sequester carbon and eliminate pollutants from the air. Cleaner air reduces risks to health problems such as asthma, which have increased tremendously in the past decade. Soil also has the ability to sequester carbon.

- ***Climate comfort control.*** Food landscapes, on their own and when combined with buildings and other city infrastructure systems, can increase the ability to provide for increased climate control and aid in the reduction of the heat island effect generated by cities. As we lose permeable surfaces that support vegetation and soil life to paved surfaces of streets, parking lots, and buildings, we raise the temperature in a city. Orchards and cultivated forests can provide shade and cooling effects to the environment.

- ***Productive edible landscapes for local community food justice.*** Parks and linear greenbelts that include edible farms or food landscapes located in underserved areas of a city should include animal pollinator management to increase biodiversity and productivity. Edible landscapes can also be used for transition zones between public/private spaces through the use of plantings such as orchards or hedgerows. Working or demonstration parks are a newly emerging urban ag typology in urban parks such as London's Mudchute, City Slicker Farms in Oakland, and the farming program associated with Havana, Cuba. The edible farms in Cuba now provide over 60 percent of the produce consumed by the city.

- ***Connection of people to nature.*** The Nature Deficit Disorder, a term coined by Richard Louv in his book *Last Child in the Woods*, is a description of the human costs of our increasing alienation from nature. First associated with children in urban environments, it is now thought to apply to adults in cities as well. Therapeutic landscape design has statistics that show that landscapes help to heal and soothe patients as well as people walking in urban environments. Food landscapes offer an excellent chance to engage and not just experience nature in the city.

- ***Human health aspects from psychological, biological to physical.*** As metrics for human health and well-being continue to be collected through new rating systems such as SITES, urban ag landscapes will prove fruitful to increasing health and well-being benefits in a city. Adding plants that produce ecological medicinal benefit, community gardens that provide for social interaction, educational opportunities for knowledge, training and mentoring, providing for multigenerational engagement, and eliminating harmful pesticides and herbicides from use in the environment, all add up to a healthier environment for people and ecosystems.

- ***Recycling and Upcycling of waste, zero net waste.*** Thinking of waste as a resource or, as Bill McDonough so famously quips in his lectures and book *Cradle to Cradle*, "waste equals food," is an important part of rethinking waste streams in a city. Upcycling a material to be reused in a new and improved way is one method to address some waste products. Designing for deconstruction and repurposing those materials to another use is another way to address waste. Green waste can be turned into soil and compost. The ultimate goal is to achieve a net zero waste goal. This includes thinking of the biological waste elements as one cycle that is an ongoing lifecycle always renewing itself back into the stream as a useful product and provide opportunities to manage the manmade waste products as another lifecycle of renewal and repurposed use.

- ■ ***Renewable energy.*** Urban ag landscapes can provide for increased energy conservation and capacity. Regenerative aspects include creating fiber fuels for machinery, nurseries, or greenhouses or land-based aquaculture facilities, addressing heating and cooling through building techniques such as green roofs and living walls, turning material outputs such as wood debris, paper products, and food waste into compost that is then added back into the food landscape as fuel. Alternative energy sources of the future could include the use of biomass to produce heat either through thermal conversion or biogas digestion. Algaculture or anaerobic digestion may be possible in the future but these sources are not yet an economical alternative.

Lifecycle Strategies for System Integration

The term *lifecycle* refers to the useful life of a product or system. In the case of urban agriculture, it also refers to a more sustainable approach where the three main systems, ecology, culture, and economy, are actively pursuing a balance towards renewable systems through synergetic connections and relationships. Thus, the lifecycle strategies for systems integration need to be tailored as a dynamic and iterative process for designing and managing the landscape on an ongoing and self-sustaining manner. Resiliency of the ecology and community is the optimum goal.

With urban agriculture landscapes, the ultimate sustainability goal is to design systems that allow for accommodating a dynamic of interdependence. Every design decision and adjustment has the potential to provide for the systems and their networks to function or not function on a more integrative level. Interconnectivity and interdependence are two of the primary ways systems work together or within themselves. Fluidness, flexibility, and cyclic responsiveness are key terms for guiding system integration. To achieve a viable sustainable food system within a city, an urban agriculture landscape is most successful when based on an integrated system approach that links natural systems with built systems to achieve a food producing landscape that will benefit the community. With an integrated systems approach, the evolution of how the systems are designed for integration into the final design output is what will support a living process.

How the systems are designed for integration into the design becomes more achievable if the system connections can be mapped. An urban ag systems matrix is a mapping tool that could be used to monitor and evaluate the potential connectivity aspects of the proposed systems. The more connections that can be achieved, and then later maintained, the more the systems network will function with a higher sustainability outcome once the project is operating. Before a project begins to integrate and link the systems within its specific scope, a working understanding of what is considered in each system is required so that the appropriate design strategies for integration can be more readily determined and harnessed.

As the design develops for the urban ag landscape, site strategies to increase system connections should be examined to determine which ones are going to be the most appropriate for the project's specific site and location. Testing the strategies at various check points as the project progresses will help to provide for a fluid, dynamic approach. Sometimes it will be hard to predict the exact outcome, but a systems matrix is a useful tool to chart them. In many cases, parts of a project may be phased over an extended time frame so charting the overall connections and the interim connections also provides a road map for the future as the project raises the funds to realize its goals.

A traditional lifecycle approach is a cradle-to-grave system. This is a finite system that encompasses a beginning, middle, and an end and does not foster systems integration since it results in abandonment, waste, or ending. For urban agriculture landscapes, a finite system is not the right choice, nor is it the sustainable choice. Food landscape typologies require an ongoing lifecycle system to ensure project success over time. In order to achieve an eco-balance of renewable systems as the project progresses or evolves, the lifecycle operations would be designed to allow for modification and evolution of its business structure, its technologies, its production math, and its human levels of input as required to maintain a self-sustaining level of existence. This includes being able to accommodate a flux of the old systems with the new and at the same time providing for systems mitigation strategies, replacing old technologies with newer, more innovative ones, and providing for a means to optimize the allocation of necessary inputs into all of the systems in order to continually integrate them into meeting the ongoing and potentially evolving objectives. This lifecycle approach can be modeled on the continuous aligning of the business and the management flows with the project's vision, mission, and objectives. It seeks to evolve the operations as the vision evolves over time in a continuum of systems fluidity.

A few examples of lifecycle site strategies that can be incorporated into the design to promote system integration or contribute to connectivity synergies in an urban agriculture landscape include:

- Stormwater quality treatment to increase health of local watershed and increase low-impact development solutions on site
- Establishing habitat corridor connections within an open space network of the community that also connects pedestrians to project site
- Designing for capturing renewable energy as an output of urban farm and food landscape operations
- Providing for locally sourced materials and products to increase local economic benefits
- Designing for on-site lifecycle waste management to increase environmental and economic benefits to local community and reduce the strain on city infrastructure capacity

Lifecycle systems strategies will be discussed further in Chapter 4. Lifecycle strategies for maintenance and operational systems and for long-range and short-range strategies for reinviting food back into cities will be discussed further in Chapters 5 and 6.

Scent of Orange, Chongqing, China

Scent of Orange is a master plan for a comprehensive new sustainable community located in Chongqing, China, along the banks of the Yangtze River (Figure 2.23). The SWA Group provided master planning services with a goal to implement the government's recent initiative, Integrated Rural Urbanization Program (IRUP), to promote development in China's rapidly growing rural areas. The project (Figure 2.24) will become a prototype for development of rural areas while conserving the agricultural resources of the Chongqing region. The plan develops a sustainable, multiuse community focused on agriculture, tourism, agricultural research, residential, and recreation uses anchored by a new research university on this historic agricultural site located within 30 minutes of Chongqing (Figure 2.25).

The Scent of Orange community is a mixed-use development with an aggressive agricultural mandate that preserves 80 percent of farm land in productive uses on the 3,200-hectare site (Figure 2.26). The Scent of Orange Master Plan is being prepared for a major produce distributor-cum-developer, which creates the unique opportunity of a developer invested in the agricultural plan as much as the development plan.

A hilly region of China, through studying the traditional response to the land (Figure 2.27), the designers discovered that a clear pattern emerged in relationship to the topography, in which buildings are placed on slopes and high points, and valleys are used for agriculture. The unique relationship between the present rural development and the site topography inspired a strategy that identifies farm home sites to be retained, and transformed into villa sites, boutique hotels, restaurants, or even relocated farmer housing (Figure 2.28). This would reinforce a physical relationship between development and the land and further defines the agricultural landscape as a primary identity of the development.

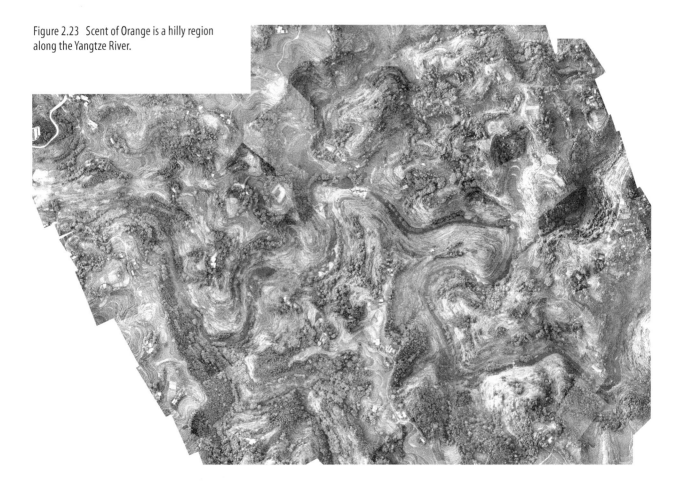

Figure 2.23 Scent of Orange is a hilly region along the Yangtze River.

Planning Strategies for Urban Food Systems 81

Figure 2.24 The project master plan integrating agriculture into the development plan.

82 Designing Urban Agriculture

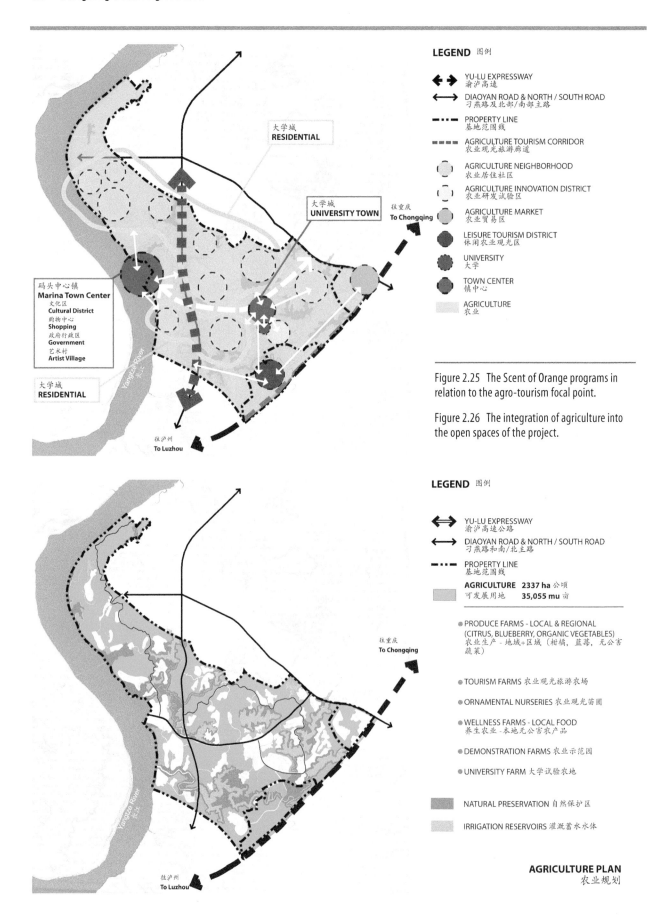

Figure 2.25 The Scent of Orange programs in relation to the agro-tourism focal point.

Figure 2.26 The integration of agriculture into the open spaces of the project.

Planning Strategies for Urban Food Systems 83

Figure 2.27 Traditional cultural hillside farming practices.

Figure 2.28 Agriculture is farmed as the primary identity of the development.

The Scent of Orange Master Plan suggests physical and economic models of development that preserve and enhance agriculture, a critical goal for the Chinese government, given social migrations and growing concerns about future food security. The project seeks to engage the challenging issues of modernizing rural China. By working within the social context of the landscape, the master plan suggests that the landscape is equally about the people who live and work on it as it is about the natural systems. By setting forth a vision of agriculture that looks forward to the future rather than the less sustainable practices of the past, Scent of Orange develops a prototype for understanding agriculture as an integral part of development rather than in opposition to development. The prototype fosters an integrative systems approach using ecosystems, social systems, and economic systems as the drivers to create a master plan that treats the food system as an integral layer to the community and region.

Design Team:	
Client:	CHIC Group Global (in Partnership with the Chongqing and Jiangjin governments)
Agriculture:	CHIC Agritech Company, Ltd.
Master Planner:	SWA Group
Water Resources:	Natural Systems International & Biohabitats

The Urban Ag Design Process Spheres

There are a number of stages and tools for integrating food landscapes into a city or community as well as the ecological benefits that are created. The process for these landscapes can follow the traditional design process, as most development projects do and cities direct projects to follow. However, that process is linear in nature and not organized around developing for system connectivity. A more cyclic design process would provide a few key differences. The differences are related to the creation of an integrated systems framework that considers how to build regenerative capabilities into the system, strives for achieving resiliency, and endeavors to achieve a sustainable lifecycle approach that adds value to the ecological, economic, and social benefits of the community.

Aspects of the urban ag design process would also consist of identifying the following:

- Potential policy hindrances at city/county level upfront
- Key stakeholders as early as possible, and thus accommodating the possibility for more intricate systems integration with the team to address the site development issues
- Funding aspects upfront
- A preliminary start-up and annual budget strategy that includes understanding the potential business plan model
- Lifecycle management commitments upfront

If one were to begin to integrate the urban agriculture planning strategies we have identified in this chapter, one must also retool the design process into a more circular or cyclic process as opposed to the linear design process that most project developments typically follow. The design process includes a cycle of spheres of connection between policy, planning, vision, synthesis, integration, lifecycle operations, and outreach and then back again to policy. All are connected and linked throughout the process, which evolves as the project development evolves. We will take a look at how the Urban Ag Design Process Spheres (Figure 2.29a and b) process work in the next chapter.

Planning Strategies for Urban Food Systems 85

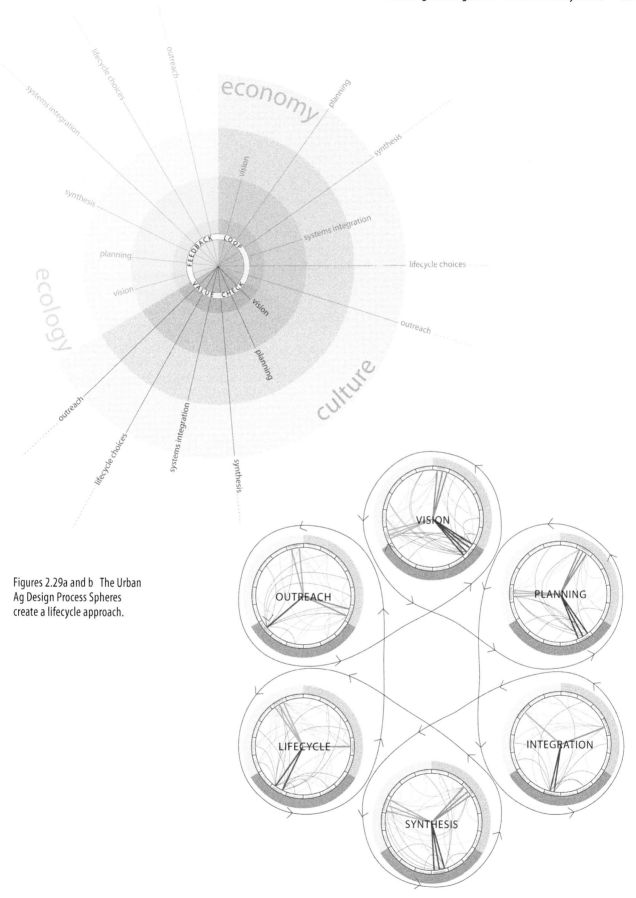

Figures 2.29a and b The Urban Ag Design Process Spheres create a lifecycle approach.

Resources

Ackoff, Russell L. *Systems Thinking for Curious Managers*. Axminster: Triarchy Press, 2010.

American Society of Landscape Architects. "How to Expand Urban Agriculture." 2010. http://dirt.asla.org/2010/01/28/how-to-expand-urban-agriculture/.

Bateson, Gregory. *Steps to an Ecology of Mind*. Chicago: University of Chicago Press, 2000.

Boyd, Susan, and Roy Chan. *Placemaking: Tools for Community Action*. Washington D.C.: CONCERN, Inc., 2002.

Butler, L., and D.M. Moronek (eds.) *Urban and Agriculture Communities: Opportunities for Common Ground*. Ames: Council for Agricultural Science and Technology, 2002.

Center for Ecoliteracy. *Big Ideas: Linking Food, Culture, Health, and the Environment*. Berkeley: Learning in the Real World, 2008.

Covey, Stephen M. R. *The Speed of Trust: The One Thing That Changes Everything*. New York: Free Press, 2006.

Holmgren, David. *Essence of Permaculture: A Summary of Permaculture Concepts and Principles*. Hepburn: Holmgren Design Services, 2012.

Kalafa, Amy. *Lunch Wars: How to Start a School Food Revolution and Win the Battle for Our Children's Health*. New York: Penguin Group, 2011.

Kelly, Kevin. *Out of Control: The New Biology of Machines, Social Systems, and the Economic World*. Jackson: Perseus Books Group, 1994.

Louv, Richard. *Last Child in the Woods: Saving Our Children from Nature-Deficit Disorder*. Chapel Hill: Algonquin Books, 2008.

Meadows, Donella H. *Thinking in Systems*. White River Junction: Chelsea Green Publishing Company, 2008.

Mollison, Bill. *Permaculture: A Designers' Manual*. Tyalgum: Tagari Publications, 1988.

The Prairie Crossing Learning Farm. "The Learning Farm." 2012. www.prairiecrossing.com/farm/learning.php.

de la Salle, Janine, and Mark Holland, eds. *Agricultural Urbanism: Handbook for Building Sustainable Food Systems in 21st Century Cities*. Winnipeg: Green Frigate Books, 2010.

Smit, Jac, Joe Nasr, and Annu Ratta. *Urban Agriculture: Food, Jobs and Sustainable Cities*. New York: United Nations Development Programme, 2001.

CHAPTER 3
Vision, Synthesis, and Form

Villa Augustus, Dordrecht, The Netherlands

Villa Augustus (Figure 3.1) is a kitchen garden, hotel, and restaurant in Dordrecht, just 15 miles away from Rotterdam in the Netherlands. The hotel is in an ornate nineteenth-century water tower located on a former derelict industrial site (Figure 3.2) that has been reclaimed and reinvigorated by the hotel development. According to its owners, it is the resort's edible and ornamental gardens that are most important to the hotel's identity and appeal (Figure 3.3).

The gardens are managed by Villa Augustus's co-founder Daan van der Have, who, along with his partners, learned about the tower in 2003. Van der Have had been working extensively on his own two-acre kitchen garden at the time, and was enamored with the experience of working and recreating with family and friends within the garden environment. When he saw the tower and its surrounding grounds and abandoned water basins, he found the perfect opportunity to create this garden experience on a grander scale for others to experience. It was following this personal revelation of van der Have's that the group decided to renovate the water tower and create a hotel.

Figure 3.1 The gardens and greenhouses of Villa Augustus in Dordrecht, the Netherlands.

Figure 3.2 The site history as a water tower influenced the design and materialism.

Figure 3.3 Edible and ornamental are integrated to create a function yet beautiful aesthetic.

The gardens not only help to define the physical space of the hotel's grounds, but are intimately connected to its operations as well. The hotel's restaurant customizes daily offerings based on the garden's produce (Figure 3.4), and the Villa Augustus chef is involved in helping to select each season's planting plan. Going far beyond a simple vegetable garden, the grounds are home to over 100 different crops, including leafy greens, herbs, and berries. An orchard on the property contains apples, pears, plums, and cherries, providing fresh fruit throughout much of the year. Ornamental flowers from perennials and edible beneficial plants are cut to decorate the hotel interior, and edible flowers are used in the restaurant. They also manage to grow cold-season vegetables in the Dutch winter through the use of a few greenhouses that border some of the gardens. The restaurant is located inside the former pump house (Figure 3.5). For a successful kitchen garden that works on all of these levels, van der Have believes that the gardener must become a little bit of a chef and the chef must become a little bit of gardener.

Figure 3.4 Harvesting produce from the garden to be served in the restaurant at Villa.

Figure 3.5 The restaurant at Villa Augustus works with the garden to determine seasonal menus.

Figure 3.6 The rooms at Villa Augustus open onto the gardens, connecting the visitor to nature.

Figure 3.7 Repurposed materials were used throughout the garden.

Figure 3.8 Visitors to Villa Augustus are invited to stroll the gardens and take in the beauty of the agrarian landscape.

The gardens were designed to serve the hotel not only via providing food for its restaurant and market but also to be a key feature of the site. The gardens blur the lines between the indoor to outdoor relationship with the hotel guest rooms, inviting guests to wander and enjoy the serenity of the productive landscapes (Figure 3.6). Occupying the spaces of the site's former water basins, the various areas in the garden each have their own unique character and design features. Though sections of the garden, with their geometric formality, allude to the Italian tradition, other areas are intentionally loose to provide contrast and respite.

Van der Have designed the grounds himself before construction began, but he and his partners understood the need to keep their intent flexible as the garden took shape. For instance, during demolition they discovered some beautiful brickwork used as fill in the old water basins. Rather than removing it, they used it to build a brick wall for the gardens. The site's other legacy materials were likewise used and repurposed when possible, carrying the vernacular of the site's past into its present and future (Figure 3.7).

The garden has been a great success for the owners. It has attracted many visitors, both domestic and foreign, who desire something much more than the traditional hotel stay. Villa Augustus offers a one-of-a-kind experience, connecting nature, architecture, history, food, and hospitality (Figure 3.8). It represents the lifestyle hotel of the future.

The Urban Ag Design Process Spheres

Urban agriculture is not currently regarded by most city planning departments as a key sustainability component of the city planning process. It has not yet been easily incorporated or grouped with the green infrastructure components of a city's systems. Until recently, urban agriculture has primarily been seen as a rural and peri-urban commodity located on the fringes of a city or outside of the city. Within the city limits urban agriculture has mainly manifested in physical form as community gardens or farmer's market enterprises. With the new wave of urban agriculture landscapes beginning to appear and the emerging national dialogue on the role that ecological urbanism and landscape urbanism must play in making our cities more sustainable, we are beginning to realize that there are so many more opportunities to capitalize on with these landscapes. Urban agricultural landscapes, or more specifically *food landscapes,* when seen through its social, ecological, and economic dimensions, offer an urban community an exciting new way to add value and prosper.

> **A key paradigm shift for city planning is to think of urban agriculture or the food landscape of the city as a prime ingredient of the green infrastructure of the city and of a city's health. Through urban agriculture policy the food system can be delineated as** *food sheds* **within a city.**

Much like a watershed defines a city's water system zones or precincts, this landscape of food can be made identifiable as a city or community's food sheds, defining the food system components within a city.

Integrating urban agriculture planning into the city is best served by retooling the design process into a more *circular* or *looped* process as opposed to the more *linear* design process that most project developments typically follow. This circular process includes the components of the linear design process but differs by rethinking the system components that connect into the overall systems network as a series of *loops* instead of *endings.* These loops connect back to the overall system through inputs and outputs within the subcomponent layers of the system. The network expands to address the system layers of these components and subcomponents. It also embraces the components found in the before-and-after phases of a linear process, thus evolving into a more complex process that reconsiders the lifecycle aspects of the system as an ongoing and self-sustaining component of the overall system. Whereas a linear process sees a project's development steps as the means to an end, ending when a project is completed, the circular process sees the project's development as a series of flowing spheres that cycle though a project's development in feedback loops. This process re-envisions project development as a *lifecycle* with the project not ending abruptly but continuing forward in its evolution of operations, management, and marketing and outreach in a more sustainable manner. It is a dynamic process rather than static one. In an abstract way, the interconnectivity of systems is much like an infinity diagram or a picture within a picture, where the picture is endlessly repeating itself into infinity. If the system components are connected in a sustainable manner, the system flow should become more fluid, circular, self-sustaining, and ever evolving. This circular lifecycle process can be diagrammed into the urban ag design process spheres, and it is the most important factor that contributes to success in the creation of urban ag landscapes.

Urban ag landscapes are living organisms that require system inputs and outputs throughout their lifecycle. These landscapes when modeled on a systems thinking approach allow for the system inputs and outputs to regulate the lifecycle flows in a self-sustaining manner. If the design process followed in creating these landscapes does not design for the entire system or subsystems to provide the ecological, social, and economic connection loops back into the community fabric or consider the ebb and flow aspects of these types of living landscapes, then success will be fleeting at best.

Figure 3.9 illustrates what the urban ag design process spheres look like. Start with the most simple broad-brush form of its spheres. Keep in mind that the process will change as system components expand or contract, depending on the project type, size,

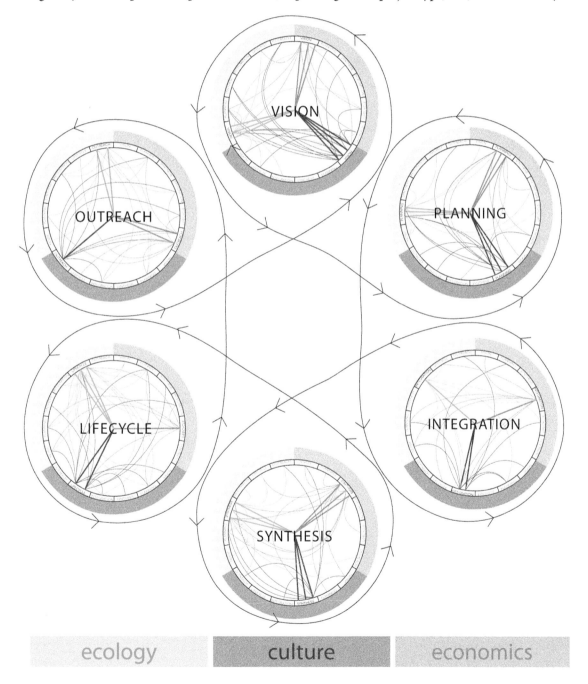

Figure 3.9 The Urban Ag Design Process Spheres diagram demonstrates how the six spheres work within the three core influences of culture, ecology, and economy.

ecosystem services, and specific community resources that are available. First, at the heart of the process are the core spheres defined by most sustainable design experts as *ecology, culture,* and *economics*. These interlocking spheres anchor this systems-based approach as a sustainable-based process. It is through the balancing and finessing of the interconnections of the three core process spheres that urban ag landscapes can begin to tackle the larger issues that plague our cities such as food security and food justice to begin to approach solutions from a more sustainable lens that is part of the culture of a sustainable city. The six overall urban ag process spheres radiate around the three core process spheres. These six spheres are: the planning sphere, vision sphere, synthesis sphere, systems integration sphere, lifecycle sphere, and outreach sphere.

The circular lifecycle process also provides for spheres within the spheres as each system component expands or contracts with resources, connections, and choices made. Key subcomponent spheres include items such as: the identification of potential policy hindrances at city or county level; key stakeholder engagement at multiple city levels; systems integration and communication; site development requirements that are designed for input and output data; identification of funding sources for both startup budgets and annual projections; the preparation of an urban ag business plan to guide the project; the branding and marketing plan; or, understanding the lifecycle management commitments required for supporting an urban ag landscape. Additionally, subcomponent spheres include the planning effort required for integrating local ecosystems into the green infrastructure of the city plan, which might even begin as a grassroots driver before actual policy changes occur.

This lifecycle process approach also includes key *value checks* along the way to test the system(s) connections that are being tapped or connected to and provide data for evaluation of the input and outputs of the systems that will continue to build the lifecycle network of the project. These value checks become the first tier of feedback loops for the project as it evolves and moves forward into a reality.

Process Spheres Synopsis

We will start with a brief description of each design process sphere before launching into more detail of each one in the following chapters.

SPHERE 1 = PLANNING & ADVOCACY... *Policy/Planning Strategies & Methodologies/Advocacy*... This sphere includes fostering and creating an urban ag policy framework within a city or community, harnessing sustainable planning methodologies, and setting a foundation for community advocacy. It is the connector sphere as it precedes and antecedes a project in a lifecycle process.

VALUE CHECK = FEEDBACK ... At various milestones between spheres, a critical component in the process is to schedule an impartial planning evaluation with the team and stakeholders. At this milestone, it would include testing the system strategies, evaluating the vision and community methodologies, and verifying that system connectivity can support a lifecycle process within the community or region.

SPHERE 2 = VISION... *Vision/Collaborative Conversations/Design Typology/Communications*... This sphere is about setting the vision narrative—identifying the mission, goals, and objectives. It includes site selection, facilitating collaborative conversations,

The potential is high for green job creation within this sphere as well as for escalating community grassroots education and advocacy efforts to mainstream consciousness and outreach. If policies are already in place, because they are relatively new and there is a perceived lack of interconnectivity between departments, the need exists for identifying a "green tape cutter" to assist the project flow through the approval and permit processes. This process sphere also sets in motion the community stakeholder communications input and output systems between city and community once policy and guidelines are established.

Planning Strategies and Methodology

This sphere includes harnessing sustainable planning methodologies and strategies detailed in Chapter 2. These strategies aid in the development of planting seeds that aid in developing the vision, the next sphere in the process for a project. The planning dialogue between design and planning professionals, local governments, and the community should focus on the need for integrated infrastructure, urban open space design, human scale agriculture, and the creation of more flexible policies that allow for food within a city. This new planning approach must consider the integration of natural systems with urban systems into city infrastructure systems. City infrastructure needs to include urban agriculture systems, or more specifically, urban food systems as an integral part of the citywide fabric. Urban agricultural concepts, especially nonconventional ones, can offer new ways to rethink city infrastructure systems to move water, energy, people, and materials that will benefit both the city and the regional systems. It also needs to focus on integrated systems thinking. With integrated systems thinking the focus is on solutions that are based on the interconnectedness of the systems as a whole unit rather than as separate independent units. The network of the systems is a vital component that functions better with integration and interdependence rather than by separation. See Chapter 2 for more detail on planning strategies and methodologies and Chapter 6 for more specifics on policy and advocacy.

The Vision Sphere

The vision narrative sets the stage for identifying the mission, goals, and objectives of agricultural landscapes, creates the big idea into physical form, and sets the framework for the core team, stakeholder, and community communications network with an emphasis on the collaboration and the creation of feedback loops. The design typology is selected and a potential site is identified that meets the vision.

Facilitating Collaborative-based Conversations

The vision process for urban ag landscapes begins with collaborative-based conversations and ideas. Experiencing collaboration in its basic form is to participate or cooperate on a joint project or task. It is seldom experienced more than as a shared event. A more transcending experience occurs when a collaboration is genuinely engaged by people who value the ideation process and each other's opinions. The collaborative conversation approach is based on the idea that better choices are made available with more voices at

Figure 3.10 The planning sphere diagram highlights the steps and stages particular to the planning sphere and how they relate to each other and the five remaining spheres.

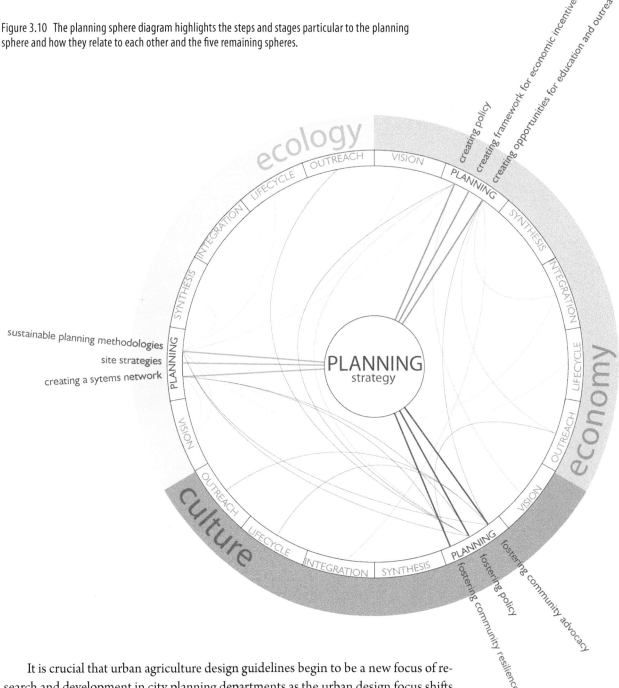

It is crucial that urban agriculture design guidelines begin to be a new focus of research and development in city planning departments as the urban design focus shifts to providing for a more sustainable future. Urban agriculture design guidelines can be used to affect policy and be created as an outgrowth of policy. Policy change may start with community grassroots advocacy rather than from internal city planning departments. The grassroots movement is currently the key momentum driver in most cities in the adoption of policy that promotes urban ag landscapes. Cities should look to these grassroot private and public enterprises and make sure that new urban ag guidelines do not prohibit the innovation and potential land use breakthroughs these projects can help to provide.

VALUE CHECK = FEEDBACK …This value check is focused on the transition from team to stakeholders as the project tracks toward going live. Before the project is constructed, lifecycle choices should be finalized with the stakeholders. At this checkpoint, if the stakeholders' budget plan includes grants and other type of funding applications, these should be monitored and confirmed in the business plan to ensure that adequate funding from the public or private realm has been obtained or is underway.

SPHERE 6 = OUTREACH & ADVOCACY…*Marketing/Education/Monitoring/Research/Advocacy* …This sphere concentrates on developing, marketing, and maintaining the brand. It considers expanding the outreach from grassroots to mainstream education. Through the development of an annual funding program outreach can foster green job creation, which includes establishing a system for monitoring and research, developing case study data through educational partners, community advocacy for urban ag programs, resource system protection, and connectivity. Project team and community stakeholder communications input and output continues and transitions to new relationships.

VALUE CHECK = FEEDBACK… This value check asks the tough questions one more time: Is the project meeting its vision, are the budget and funding mechanisms sustainable, is the outreach being achieved, are the systems connections and integration design actually occurring and achieving the sustainable lifecycle goals? At this time, the project's data are also at a point where it circles back into the sphere of planning and advocacy.

The Planning Sphere

The planning process sphere (Figure 3.10) sets the stage for the success of urban ag landscapes. Policy and advocacy are the vehicles to support urban agriculture within a city.

Planning, Policy, and Advocacy

Planning is the connector sphere between the vision and outreach spheres of the urban ag process as it comes both before and after a project experiences its lifecycle evolution. Without adequate policies in place, many projects cannot actually happen. When designing urban ag landscapes it is important to identify upfront any potential hindrances at city or county level that would defeat a project from moving forward. While some cities such as New York, San Francisco, Chicago, Detroit, and Baltimore have made great strides in allowing and inviting urban ag landscapes to occur within the city limits, many cities still have old laws, ordinances, and policies that make it harder and even impossible to build them within the city limits. From the urban design perspective, this sphere supports the creation of sustainable city planning strategies that include urban agriculture. It should promote city department interconnectivity (not a simple task given the linear hierarchy of most planning departments), create planning guidelines that will aid in establishing resource system protection and connectivity at community, city and regional levels, and create a structure that is supportive to cities and communities in identifying the synergy between city/state/federal policies.

identifying the stakeholders, and identifying potential funding opportunities. Key factors in this stage include identifying the urban ag design typology, selecting the core team, and setting up a communications network that adds to community engagement. A framework for establishing team, community, and stakeholder communications input and output is begun to establish how feedback loops will be integrated into the overall process.

VALUE CHECK = FEEDBACK … At this milestone between spheres, a critical component in the process is to schedule an impartial vision evaluation with the team and stakeholders. At this milestone it would include testing the proposed system strategies for the selected typology, evaluating the vision and community communications methodologies, and that feedback loops have been set in place.

SPHERE 3 = SYNTHESIS … *Synthesis/Idea and Systems Framework/Business Roadmap* … This sphere begins with site and community systems resource evaluation and the conceptualization of the vision or idea into physical form. It considers factors of scale investment, developing the system schematics and connections, and synthesizing the vision with the site and community systems framework to develop the roadmap for connectivity including fostering integration within a larger project if urban ag is a component of a larger project vision. Developing the preliminary urban ag business plan is important at this stage to provide a roadmap for development. Project team and community stakeholder communications input and output is ongoing for the feedback loops.

VALUE CHECK = FEEDBACK … At this milestone between spheres, a critical component in the process is to schedule an impartial project synthesis evaluation with the team and stakeholders. This task evaluates the solidness of the data synthesis and tests the preliminary systems framework plan to verify that system connectivity can support a lifecycle process within the community or region. It evaluates the preliminary budget plan and evaluates whether the systems framework plan has traction with local community and jurisdictions.

SPHERE 4 = SYSTEMS INTEGRATION … *Integration/Connections/Interdependence* … This sphere is focused on developing the design and ecosystem components, including the ongoing testing of the systems and strategies through detail development and exploring new technologies, and the evolution of the systems integration into final design output for appropriate permits, codes, and approvals. Systems integration should verify the connections to the community wide and citywide systems. Refinement of the urban ag business plan and budget occurs. Project team and community stakeholder communications input and output is ongoing for the feedback loops.

VALUE CHECK = FEEDBACK … This value check in the process is focused on an impartial project systems integration evaluation with the team and stakeholders**.** This feedback loop consists of retesting the systemwide connections, evaluating whether the proposed business plan is tracking with the vision, and verifying the connections to other systems within community and region have been set in place.

SPHERE 5 = LIFECYCLE CHOICES … *Maintenance/Management/Operations/Funding* … This sphere begins with the outlining of the future management infrastructure framework for maintenance and operations. It develops the lifecycle budget and phasing opportunities, creates a plan for harnessing green job potential, and finalizes the urban ag business plan before the project goes live and sets up the transition from team to client/community oversight. Project team and community stakeholder communications input and output continues. Preliminary outreach begins.

the table. It is inclusion oriented and values diverse perspectives and thinking through active listening. Facilitating collaborative conversations can help to identify the project's potential stakeholders and help them to work together to solve problems, craft visions, or create results-oriented action plans. This type of conversation has the ability to pinpoint ways to discover the specific needs of a project.

The vision may start with a singular source point but may also be group driven, depending on the context and scope of the project. It may vary with the size and complexity of the idea. Additionally, it may depend on whether the project is in the private or the public sector. The design process for a private project could require a vision on a more individual level with fewer stakeholders involved in the initial stage, whereas a public project might require a vision to be developed on a communitywide scale that includes a fair amount of stakeholder discussion to craft the vision. Either way, facilitating collaborative conversations is useful throughout the process, whether it is with clients, community stakeholders, neighborhood groups, school districts, city agencies and commissions, or potential donors and grant foundations.

At the start of the vision process, whether it is between only a designer and client or between many members of a community, the beginning conversations start with a *needs and resource* analysis to identify what the community has, what it needs, and how to connect the two. This analysis might kick off with a subjective understanding of the issues but needs to be based on real facts and figures before dialogue occurs. Facts become the basis for the vision metrics or motivating metrics underscoring the need that aids in inspiring a vision. The facts should be relevant to the idea and the urban ag typology that forms the core of the vision. Once the facts are in hand, then multivalent conversations can ensue with the stakeholders. Market research is helpful to identify the stakeholders when they are not yet known.

Any time a change happens to us humans, our brain thinks that it is something to run from. The amygdala is the part of the brain that is responsible for our fight or flight responses. It recognizes patterns in our day-to-day life, and it likes to keep us near what is familiar because it has been proven to be safe. Anything new is automatically a trigger of potential danger that needs further investigation. This takes time and is a process of incremental exploration, trust in whom or what is delivering that change, and a chance to influence that change in order to feel a sense of ownership to it, and that it will be something that adds benefit to our lives. Also, when things affect us in our own community we like to be heard in our expression of what we want in a decision of changes to come. It is important to follow fair community processes that genuinely help the community voice be heard and acknowledged in the design of the changes. The typical process of a public hearing and input process of standing at a microphone and talking to a panel to decide, or the process of writing to policy makers, or trying to get access to a design and planning firm can fall short in this regard.

Facilitators all over the world are now using a suite of what are called *social technologies* to move large groups of stakeholders through engagement processes that reduce the barriers typically seen in formal processes that protect the decision makers. These social technologies can bring people together to tap into the collective intelligence, to find innovation that happens when differing expertise talk together, and allow the power structures to work in a healthy empowering way with the community voice. When the community voice is heard in the beginning of the design and planning stages, the projects going forward can be more successful due to the buy in and acceptance from the community because they trust and know the project to be created by them. Also, relationships are built between stakeholders so there is enough trust in the project decision makers that they will make the right choices. The newness of the project that our human brains do not like will not be seen as much as a threat to our safety.

Once the barrier of fear is overcome, the community can then move into creativity and collaboration toward a sustainable food system that they feel proud ownership of. A key part of the sustainable success of the food city system to maintain itself through helpful policy, community help working on the project, and awareness of the project for economic viability is the sales generated by the system.

Many stakeholder processes can work to achieve community buy-in, trust building, and long-term willingness to work and purchase from the food city system. Many practitioners of these techniques recommend using a facilitator. It takes a good amount of expertise to design a series of collaborative conversations with the stakeholders of projects. Reaching out to experienced facilitators to design and host these conversations is well worth the investment to reduce the possible headaches of fear and resistance later in the project. Each of the following processes has a network of experienced facilitators to help in thinking through which technique is best for what situation, and how to design and manage the process.

World Café

Collaborative conversation hosting is what the World Café technique is about. These collaborative conversations aim at allowing for the collective intelligence to work through the challenges and to identify the opportunities that are held in common by the stakeholders or community members (Figure 3.11). Using social technologies can aid in allowing for the community to be able to buy in to the design and processes that will be created by the urban ag landscape and food shed opportunities that would be considered. These types of conversations have the ability to develop or enhance long-term feelings of ownership and positive quality of life and stewardship of the food landscape.

The World Café is a process based on a few key concepts. People in the community already have the innovative answers to tough problems. The wisdom is already in the room, as the facilitators like to say. If we cross-pollinate ideas with everyone in the room and build on ideas, we will create things we could not have alone. And the simple process of sitting together at small tables to explore together is a very old and familiar process of being with others to discuss any range of topics that are meaningful to us. Very important is a very well done invitation put out by a trusted, nonbiased person or organization to ensure a diverse and representative group of the community. People may find the current communications processes used as inaccessible and hard to engage with. Making an invitation up front that clearly states a different process will be used can increase attendance, accessibility to decision makers, and more voices being heard. People like having a voice in their community and they can be told that this process can help with that.

Here is a snapshot of how this process works. Community members get together to explore an important question about a project or issue in their community. The participants are then reminded of the cafe etiquette for a successful meaningful conversation. Most cafes then go three rounds of 20 to 30 minutes each. After each round, everyone

Figure 3.11 Community members collaborating at a World Café event.

switches tables trying to sit with people they have not already sat with and that are different than themselves in some way. During the three rounds, people usually draw on paper supplied to the tables to capture ideas and notes for the design and planning teams to use later. Also, each table can be designated a recorder to ensure that all the great ideas are captured.

This process is great for capturing ideas quickly, identifying patterns in a community of diverse people, building relationships, and bringing people of different power levels to work as peers on an issue. There is really no limit to the size you can do—they have happened with 12 people or 10,000 people. All you need is space, etiquette, a question, notetakers, and sitting together in groups of at least four. Of course, a skilled facilitator is key to invite people to this shared conversation.

Community Workshops

Community workshops are the best method to create a forum for a shared public dialogue on the topic when the community is the stakeholder group. The workshop is a community meeting with the purpose to bring together a cross section of viewpoints to exchange ideas and information about the topic, in this case the vision for the urban ag project. The factors for success include the following items: It needs to be planned for, set up early, have a clear goal and objective, have an agenda, have an effective way to invite participation, be publicized, have materials available that engage the conversation, and be led by a collaborative facilitator with a plan for follow-up information and identification of next steps.

Facilitating community conversations takes active listening skills. There are also many professional facilitators available to lead the workshops, such as *Collaborative Conversations* based in the San Francisco Bay area and founded by Ken Homer, who after a decade of collaboration with the founders at World Café founded Collaborative Conversations to offer facilitation and teach programs about facilitating.

Community workshops are an important tool to use when designing urban food systems to ensure the entire system of food is engaged in the design and execution of the system. Each community has a different set of stakeholders in many parts of the system:

politics, landowners, house owners, neighbors, rich, poor, for profits, nonprofits, volunteers, markets, growers, processors, sellers, restaurants, schools, and any number of other influential people in those sectors. Community workshops should be signed, considering what is needed for the relationships between these stakeholders so the relationships are constructive, trusting, and ultimately collaborative for innovation toward creating a new food system. Community workshops can build these relationships while simultaneously working toward designing, planning, and building the new food system.

Charettes

Similar to a community workshop charettes are a design term that encourages a collaborative brainstorming session to achieve its results. The same factors for success listed for community workshops apply here as well. When used wisely, charettes are an effective tool that can facilitate a consensus-building conversation for a diverse group of stakeholders or community organizations. Charettes need to be focused on building a trusting environment where all opinions are welcome and criticism of ideas is not welcome or acceptable. They are best when allowing for free-flowing conversations to generate tons of ideas that can then be studied more thoroughly at the next stage of evaluation. Similar to community workshops, charettes can build these relationships while simultaneously working toward designing, planning, and building the new food system.

A wonderful resource of tools for shaping or leading collaborative conversations is *The Change Handbook: The Definitive Resource on Today's Best Methods for Engaging Whole Systems*. This book is the definitive resource on some of the best methods that can be used for engaging people and whole systems thinking for organizations and community change. The book includes 61 collaborative change methods, including social technologies for facilitating collaborative design conversations with stakeholders such as Real Time Strategic Change, Open Space Technology, Community Needs Assessment, and others, plus it provides advice on when to use each one.

This is a great reference book to help identify social technologies and processes that fit your specific situation. It outlines the processes, but also why you would use them by describing the outcomes that the process is designed to achieve. It might be community buy-in, a to-do list from a complex project, idea generation, relationship building, or whatever you might need. The book has a huge range of options, and all of the techniques have been proven over many years to be effective. It also gives a bit of history so you can find a network of people practicing the techniques to enable you to find help or a suitable consultant for your project. This reference is highly respected as a go-to list by experts in the social technologies field.

One example in the book is on Seattle's Educational Network and examines the manner in which a number of initiatives such as the Local Food Action Initiative were created in order to develop frameworks focused on policy, actions, and procedures between community, districts, and city departments. The Local Food Action Initiative was a resolution aimed at holding departments accountable to create reports to inform a Food Policy Action Plan. However, as an initiative it did not hold anyone accountable to regulatory or legal actions. This initiative, by fostering collaboration and creating a framework for organizations and departments to follow, was a tool that could be used to create the means for other enforceable government actions to be more focused on local food initiatives that would benefit the community.

The Vision Narrative

Without vision it would be a mundane world. The vision sphere (Figure 3.12) is about harnessing the excitement, the passion, the anticipation, and the imagining of the future or desired outcome. To create a vision for urban ag landscape projects, programs or enterprises, it takes a perceptive nature and a creative spirit. The *vision* is what the project wants to become or embody—or in other words, its sense of being. At the core of the vision development process is the creation of the *mission statement.* The mission statement is a narrative that establishes the core values and builds the framework of the project by setting its *goals and objectives.* The mission narrative should be a clear and succinct statement about the purpose of the project and should represent the broadest perspective of

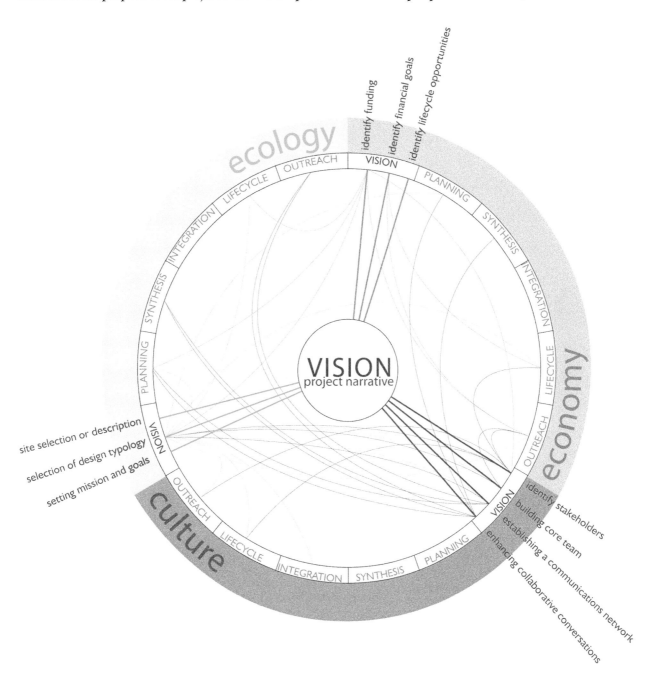

Figure 3.12 The vision sphere diagram highlights the steps and stages particular to the vision sphere and how they relate to each other and the five remaining spheres.

the project's mission in order to provide a sense of direction and guide decision making throughout the project's lifetime. It should incorporate socially meaningful and measurable criteria to provide the framework from which the project's strategies are created.

An example of how an urban ag project's vision and its mission statement are established and refined follows. This example from the Organic Edible Garden & Outdoor Kitchen at Miller Creek Middle School in San Rafael, California, began as a six-page proposal written by a sixth-grade student, Gabrielle Scharlach, that was presented to the school's principal, Greg Johnson, in January 2010. The proposal included a vision for the proposed landscape, two potential site locations on campus to select from, and case study examples that supported the proposal that included the Edible Schoolyard at Martin Luther King Jr. Middle School, begun by Alice Waters in Berkeley, California, and the White House Garden started by First Lady Michelle Obama.

Here is the vision, as first written by Gabby Scharlach, the original visionary sixth-grade student who came up with the garden idea for her middle school:

The concept of the garden is to create a healthy living space used for education and health. The freshly grown food will be used in the lunch and snack program and the outdoor kitchen will be a place for chefs to come and share their cooking skills with students. The ecosystem is the perfect place for outdoor education and the garden will beautify the school and help in fighting childhood obesity and climate change.

The principal was excited by the proposal and pulled together a group of key stakeholders, both from the school and from the parent community to form the Advisory Committee to guide the process. He also was the person who found potential organizations and people who might become the founding donors and supporters of the garden within the greater community. The formal mission statement written by the Advisory Committee, who included Gabby as founder and student representative, the school's principal, two teachers, and two parents, follows:

The Miller Creek Outdoor Edible Classroom and Kitchen is to provide Miller Creek students with an "experiential" learning environment that connects various aspects of organic gardening, basic nutrition, and culinary awareness/skills. In addition, the garden will provide a place where students can study sustainability, slow food concepts, and a "farm to table" philosophy.

The mission statement then proceeds to lay out a vision for how the edible garden would function and be operated:

The garden will operate all year and be funded by private donations and grants. It will be run by a garden coordinator. Once the garden is operational summer programs will be added to enhance the educational outreach and maintain the garden's harvest. There will be a garden Advisory Board who will set the annual program goals, approve the annual budget, plan fund raising events and work with garden coordinator. The garden coordinator will become a fulltime position once the garden is fully phased and programs developed require that need.

The garden will be built in phases. Phase 1 began late July 2010 and is expected to be complete by end of October 2010 for its first fall planting.

This garden will be 100 percent privately funded in order not to put additional financial strain on the school and the district. The garden as an outdoor classroom adds a tremendous value to Miller Creek Middle School.

Once the mission statement has been formulated and agreed to by the stakeholders, the goals and objectives then provide the means and criteria that will be used to guide the project. The goals should outline broad, general intentions, while the objectives are more precise and provide measurable criteria to achieve the goals.

An example of sample goals and objectives for the edible garden at Miller Creek Middle School follow:

Sample goal: "The edible classroom will provide students with an experiential outdoor learning environment that connects with organic gardening, basic nutrition and culinary awareness."

Sample objective: "The edible classroom will promote physical activity through garden activities and teach nutritional awareness through curriculum that is tied to state standards to better provide students with the ability to make informed decisions about healthier eating and healthier habits through hands on learning and doing."

It is important to document the vision, mission, and goals of the project at the start of the design process. This information should be collected and written as the basis for the project's preliminary Urban Ag Business Plan, also known as the Edible Business Plan. These statements will not only guide the process but also are the fundamental tool in providing direction and decision-making criteria throughout the life of a project. Sometimes, the location site of the potential project is known at the start of the vision process. Oftentimes all that is known in the beginning is that a site needs to be found to fulfill the vision and mission of the proposal. Either way, the site location conversation, known or unknown, the funding sources, the construction methods, the programs, the operations—all of these elements are included in the business plan and all of them will become part of the collaborative conversation that ensues throughout the process as the project develops.

Excerpt of Preliminary Garden Goals from the Miller Creek Edible Business Plan:

1. Provide an experiential learning environment that connects with organic gardening, basic nutrition and culinary awareness.
2. Expose students to hands-on environmental learning.
3. Enhance the curriculum by connecting it to the natural world.
4. Provide students with the opportunity to grow and eat fresh produce.
5. Build a school-based ecosystem where there was none.
6. Offer parents an opportunity to engage with the school community.
7. Develop a program that will sustain itself from year to year.
8. Develop a connection with local farms, food harvesting banks, and local chefs who will add to our school curriculum and fundraising events.
9. Make the school site more attractive in the rear of the school, which is currently neglected.
10. As an organic garden, certain sustainable practices will need to be followed so that the garden can become certified.

Miller Creek Edible Garden and Outdoor Kitchen, Marin County, California

The idea for the edible garden (Figure 3.13) at the Miller Creek Middle School in Marin County, California, came from one of the school's own students who wanted to join fight against food-related illness and obesity, while simultaneously lessening the environmental impact of herself and her classmates (Figure 3.14). The effort was partly inspired by the campaign of First Lady Michelle Obama to combat childhood obesity, as well as the organic garden she planted on White House grounds. Further inspiration came from the work of organic gardening and healthy eating advocate Alice Waters and her Edible Schoolyard Project at the Martin Luther King Middle School located in Berkeley, California.

Figure 3.13 Young urban farmers getting their hands dirty during a planting day at the Miller Creek Edible Garden.

Figure 3.14 Perspective sketch of the garden and its outdoor classroom ecosystem zones.

Figure 3.15 The vertical garden uses woolly pockets along a bare classroom wall to expand the amount of square footage for edible plants and create beauty.

Figure 3.16 Illustrative site plan identifying the sustainable building blocks and educational program elements of the Miller Creek Edible Garden.

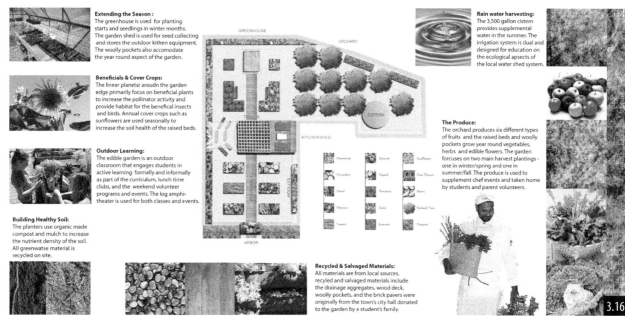

The 4,500-square-foot organic garden (Figure 3.15) uses no pesticides, and contains a 3,500-gallon cistern to harvest rainwater and a recycled irrigation water system. The garden contains an outdoor movable kitchen for cooking classes and other special events. All of teachers and students are invited to use the garden.

Highlights of the edible landscape include:

- 2,000 square feet of raised planters for herbs and vegetables (Figure 3.16)
- Living wall for herbs and micro greens along vertical face of classroom
- Rainwater cistern
- Recycled irrigation system

- Outdoor classroom-sized amphitheater (Figure 3.17)
- Garden shed for storage and roll-out kitchen (Figure 3.18)
- Greenhouse for seedlings and starts
- Produce taken home by students and their families

The school's curriculum takes advantage of the garden, especially with earth science classes (Figure 3.19).

Figure 3.17 Community educational event in the amphitheatre at the grand opening of the garden.

Figure 3.18. Cooking demonstration in the outdoor kitchen with Chef Seann Pridmore of the Epicurean Group.

Figure 3.19 Events in the garden create educational opportunities for students to teach and interact with other students.

Principal Johnson, also a believer in the "farm to table" philosophy and slow food movement, was so supportive of student garden champion Gabrielle Scharlach's proposal that within days she was asked to present her idea to the school site council, the school district board, the home and school club, and various community nonprofit organizations. She successfully raised almost $30,000 in the next three months through private funding sources excited by her idea. Because of the current economic times that our schools are in and the cutbacks that they have to face, the entire funding for the project came through donations and grants. The continuation of the garden will also come from public- and private-sector grants. This enables the garden to be totally off the financial responsibility of the school and district. Because of her leadership in launching this project, Gabby also received the 2010 Marin Youth Activist award sponsored by Senator Mark Leno's office.

The initial seed money officially jump-started the development process moving forward, and the Phase 1 garden construction by parent, student, and community volunteers was launched mid-summer 2010. The first garden-planting milestone by the students and teachers was reached the week right before the Thanksgiving break. It consisted of a diversity of cool-season winter harvest vegetables, herbs, and beneficial plants.

Principal Johnson notes that the edible garden as an outdoor classroom will add a tremendous value and create a sustainable layer of curriculum enrichment to the middle school. The overall goal of the outdoor classroom is to provide the Miller Creek students with an "experiential" learning environment that connects various aspects of organic gardening, basic nutrition, and culinary awareness/skills. In addition, the garden is to provide a place where students can study life science and environmental sustainability concepts, slow food concepts, and a "farm to table" philosophy. In Phase 2, the project is planning both solar and wind turbine power to become totally off the grid, along with the shade trellis and more development of the outdoor kitchen.

Design Team:

Garden Vision:	Gabrielle Scharlach, Miller Creek Middle School student
Landscape Architect Advisor:	April Philips Design Works
Miller Creek Middle School Principal:	Greg Johnson
Garden Coordinator:	Katie Dwyer

Identifying Funding Opportunities and Selecting a Site

It is important during the visioning stage of a project that the potential financing opportunities and a site location be planned for or identified upfront. These items can be known entities that a stakeholder or vision core committee has identified, or they can be identified as part of the vision narrative that describes the desired plan intent for the project's financing and location. Sometimes a vision is created based on a site that has inspired the vision that is both suitable and available for the envisioned project. Other times the exact site may not be known but the vision narrative outlines the goals and objectives of the site that is to be selected for the project, as well as identifying the goals and objectives of the potential funding sources.

In the case of the Miller Creek School's edible garden and outdoor kitchen, the project vision proposal written by Gabby Scharlach identified two potential sites on the school grounds that could be suitable for the garden and presented the two options to the school district board to receive their permission for the project and the preferred location. Simultaneously, the principal of the school identified two potential founding donors who believed in the project vision. Having founding donors is crucial in providing the initial momentum for the vision to move forward into a reality.

Selecting the Vision Team Core Committee

Part of the vision process is to select a design team, committees, subcommittees, and/or advisory panels that will help to develop the project into more detail in the following stages of the design process. An urban ag design team composition should be selected with regards to promoting biodiversity socially, economically, and ecologically. In other words, depending on the project's complexity, it should include those experts specific to meeting the project's needs, the community's needs, and the site's potential. These experts could include local professionals such as landscape architects, horticulturists, interested residents, or neighbors, restaurants and chefs, public agencies for land permitting, local farmers, nonprofit educational and health organizations, entrepreneurs, master gardeners, and more. It really depends on the complexity of the ag landscape as to what the team composition should be.

Our firm and others have also experienced starting projects that did not include urban agricultural landscapes in their original programs but then the urban ag component grew out of discussions with the client and/or community groups as the project went through the city planning and design process, which typically included social technologies such as community workshops or charettes. For these projects, experts were added to the team as required once the idea took enough shape to be included into the final program. As an example, on one of our firm's high-density residential projects, 2001 Market Street in San Francisco, California, when we were in the schematic design phase during one of the team's LEED and SITES Pilot discussions regarding the sustainable landscape, we began to describe how we could maximize biodiversity and human health through the inclusion of a community garden. The idea was well received by the client and the team of stakeholders so we began in earnest to develop ideas of how to incorporate it into the project. The initial ideas included an edible herb wall, community garden planters for the tenants on the third-floor podium level, an edible landscape on the street in the historic garden, or the addition of an urban farming component to the roof—all of which potentially could be seen as an amenity for the community and would fulfill many of the project's sustainable goals. In this case, the urban ag component was researched and championed as an idea by the landscape architect and then embraced by the client and team. A proposal was written up to identify the potential impacts this decision would bring to the project so that it could be evaluated and vetted as the project developed. A preliminary business outline of potential ways the urban ag landscape could be maintained and operated was also developed at this stage to ensure that this framework could be built into the project as part of a lifecycle initiative. Since the project was set up as minimum LEED Gold and had been selected as a SITES Pilot project early on in its development phases, the sustainable goals established at the beginning of the project acted as a framework for these ideas and others to be evaluated and nurtured into fruition.

2001 Market Street, San Francisco, California

The mixed-use residential project (Figure 3.20) at 2001 Market Street will transform an underutilized urban site into a high-quality, high-density, mixed-use project with convenient access to public transit in one of the most "transit rich" streets in San Francisco. Located at one of the most visible corner parcels along San Francisco's symbolic Main Street, the one-acre project will replace the currently vacant S&C Ford auto dealership with a 30,000-square-foot Whole Foods Market on the ground floor and 80 residential condominiums above the store. Whole Foods Market is the nation's leading retailer of natural and organic foods and will anchor the project. The project's key focus is on public transit, neighborhood serving retail and enhancing the urban realm by creating vibrant retail space and inviting pedestrian gathering spaces. The overall goal is to provide "green" urban living that incorporates sustainable design and thoughtfully considered materials and features that aid in the performance of both building and landscape systems.

Figure 3.20 Illustrative plan for Market Street development with urban agricultural elements for residents, San Francisco, California.

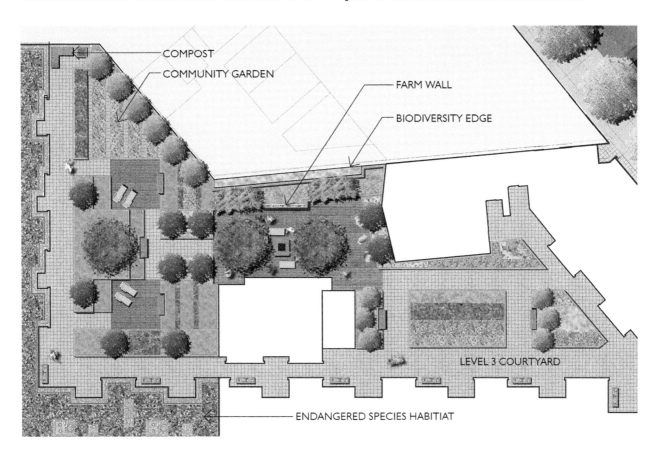

Figure 3.21 Urban agriculture brought into the podium level as an aesthetic element.

The private realm above is tied to the urban realm below through its integrated landscape systems. At the roof levels are a series of extensive and intensive biodiversity gardens. Endangered-species habitat gardens are on the fourth-level roofs totaling 8,000 square feet. The third-level roof landscape totaling 11,000 square feet is the main residential common area for community gathering, community urban agriculture, recreation, and entertaining. The community urban agriculture garden (Figure 3.21) component is approximately 650 square feet. The landscape design concept on the community gathering level roof terrace is to create a place that is a refuge and sanctuary through alfresco living. The urban agriculture component was an important aspect in developing a sense of community for the project.

The project anticipates a minimum goal of LEED Gold and the developer understands that the urban landscape is an important part of the project's success. The client is excited about the project's potential in setting the bar in San Francisco for a truly integrative urban project that would be both LEED and SITES certified as it was selected as one of the Sustainable Sites Initiative Pilot Projects. The potential for profiling substantive urban ecosystem services is high for this project. Landscape features include water catchment, living walls, urban agriculture and edible landscapes, community center, endangered species habitat for the San Bruno and Mission Blue Butterflies, stormwater treatment, traffic calming, art and educational interpretation, and a zero waste recycling system.

Water conservation is one of the key resource goals, and the team plans to go beyond the 75 percent potable water-reduction credit.

Building community is an important development goal for both the private common areas and the public open space areas. The project is targeting to achieve the majority of the SITES health and well-being credits and hope to demonstrate their real value in the urban realm. Integrating this aspect into the design framework, Whole Foods Market will create approximately 170 jobs for local residents and supports local nonprofits in the community. This project will have a "green positive" economic impact to the neighborhood.

The Integrated Landscape Systems

1. Water in the city:
   - Understand the value of one drop.
   - Consider restorative power of water and cooling effect.
   - Stormwater management principles include managing all water that falls on the site within the site.
   - Water conservation—utilize rainwater collection for irrigation and water-wise plants/drought tolerant plants to reduce water consumption (climate responsive).
   - Explore the use of a graywater system to be expanded to include low-flow toilets.
   - Use rainwater harvesting for all roofs for irrigation water.
   - Use flow-through planters or infiltration planters for stormwater cleansing at roof level and street level.
   - Green roofs and surfaces aid in managing water runoff.
   - Smart ET weather-based timer for irrigation plus subsurface microline system will add to amount of water conserved.
   - Green walls add in permeable surface and ability to move water through site.
   - Add permeable surfaces to streetscape.

2. Air Quality:
   - Plant trees to filter carbon dioxide.
   - Add an air filtration system within building.
   - Vegetation and canopy surfaces add to amount of carbon dioxide filtering.

3. Vegetation, Soil, and Habitat:
   - Intensive and extensive roof gardens will provide habitat for endangered species such as Mission Blue and San Bruno butterflies.
   - Promote living, healthy soil on roofs and in streetscape.
   - Use vegetation for beauty, food, climate control, and habitat.
   - Celebrate nature in the city.

4. Food Harvesting on Green Roof Level 3:
   - Add a community garden for use by tenants.
   - Vertical wall food garden can be used by neighbors.
   - Eliminate pesticides—use IPM (integrated pest management) for maintenance.
   - On-site compost for green waste.
   - Collect green waste.
   - Have a zero waste goal—perhaps an internal system that separates trash, recyclables, compost, and green waste.

5. Materials:
   - Eco sensitive—natural, renewable
   - Local
   - FSC-certified sustainably harvested lumber
   - Low-VOC cement
   - Other

6. Health in the City and Human Well Being:
   - Create a gathering place for community spirit, and places for meditation and refuge.
   - Provide a shared community gardening experience.
   - Exposure to nature—celebrate, experience, interact, admire, connect, and restore.
   - Provide recreation opportunities both active and passive.
   - Provide an accessible link to public transit to connect into urban fabric.
   - Cultivating biodiversity adds to human connections with nature in the city.
   - Places for ceremonial opportunities
   - Proxemic—include scale and perception.
   - Acknowledge physiological benefits and interaction to build "community."

7. Economics in the City (Balance):
   - Understand and educate how ecosystem services add value.
   - Include food production and consider creation of a green job training program or restaurant connection to garden.
   - Make the project self-sustaining and give back to the energy grid.
   - Create green infrastructure—i.e., green streets.
   - As part of the Market/Octavia district—project should add to neighborhood value o fits context.
   - Tap into the green real estate branding market—assess the green viability and walk the talk.

Design Team:	
Client:	Prado Group, Inc.
Project Architect:	BAR Architects
Collaborating Architect:	William McDonough + Partners
Landscape Architect:	April Philips Design Works
Structural Engineer:	Tipping Mar
Civil Engineer:	BKF
General Contractor:	Webcor
Lighting Designer:	Lumenworks, Inc.

Establishing the Communications Network

The type of communications that will work best for the project will depend on the project type, the scale, the number of stakeholders, the milestone timeline, and ongoing determination of what tasks need to be accomplished as the project moves forward. The communications network can be very simple or highly complex. It should also be thought of as an item that will evolve as the project's needs and the team evolves. On small projects, email communications between the core group and the establishment of a monthly or bimonthly pattern for advisory meetings with the stakeholder group works well. A website or blog site is also an effective tool for keeping stakeholders and core team aware of the process and milestones that have been planned. On larger projects, where stakeholders need to be kept in the loop as decisions are made and outreach is the main goal, a list serv or Internet forum such as Google Documents or Dropbox is a useful tool. Google groups and LinkedIn groups are other examples where a team has used existing Internet systems to communicate internally and externally to keep the communication open and transparent for all entities. The more transparency and ease there is in the communication protocols between all participants, the greater the possibility there is for an effective communications network to occur. When a communication network is developed to be piecemeal or function on a "need to know" basis, typically a project suffers with inefficiency and has a greater chance of mistakes or for participants to continually reinvent the wheel.

The Synthesis Sphere

Synthesis is about site analysis and community systems and resources evaluation. The synthesis sphere (Figure 3.22) is about providing a roadmap for connectivity, integration, and project development. It establishes a framework that provides for the inputs and outputs of feedback loops in a meaningful way. Feedback loops are the tools used to analyze the data being synthesized.

■ **Synthesizing economics:** Creating a preliminary business plan outline for the urban ag component of a project is both an economic analysis tool and an outreach tool for education and marketing. Without understanding the project's potential overall budget needs and the value-driven return on investment that is envisioned, a project will not have a clear enough roadmap to tackle its own sustainable growth and development.

Synthesizing the Design Concept and Preliminary Systems into Sustainable Framework Plans

In Chapter 2, we looked at how interconnectivity and interdependence are two of the key ways systems can work together or within themselves. Interconnectivity is based on reciprocity and dynamic equilibrium. It implies that the parts of a system are difficult to analyze on their own since a system is more than the sum of its parts. Interconnectivity is the first layer of system performance. In the design phase we begin to look at each of the systems within a project individually such as water or vegetation and connect them within themselves to insure a well designed and interconnected system. Interdependence is based on a dynamic that is responsible to others that share a common set of principles. Thus, interdependence weaves together the systems as a network acknowledging that each system is self-reliant while at the same time responsible to the other. With urban ag landscapes the ultimate sustainable goal is to design an interdependence dynamic to the systems. This would create a fluidity that allows for a flexible and cyclic-oriented network of systems that function on a more integrative level capable of becoming more self-sustaining or regenerative in nature. Reaching a regenerative level would be the most sustainable goal to be achieved.

With this in mind, the first step in developing the initial design and ecosystem components identified on the framework plan is to develop each system component's connections individually and then develop the connections to the other system components. This step requires research, understanding, and integration of the available technologies to design the components to function on a systemwide level. Research of local codes, guidelines, and policies that affect urban ag decision making aspects is also a key step in this part of the process. If a sustainable rating system such as LEED or SITES is going to be used on the project, this would be the time to identify those components of the systems that would be integrated to meet the additional sustainable goals of the systems.

Once the project's systems are understood, they can be drawn onto site plans to illustrate their individual connections before overlaying them to refine and synthesize the connections between the systems. Sometimes the connections are not clear and are easier to be seen in a matrix format or in a cartoon 3D diagram that maps the connections of the network. Developing the design components and strategies into preliminary/schematic plans and details will be described in more detail in Chapter 4 in Systems Integration.

Design Aesthetics in Urban Agriculture Landscapes

Aesthetics is always a subject that comes up when designing urban ag landscapes. There are elements in a food landscape, such as storage, waste management, and seasonality,

overlay these with identified economic analysis parameters. This map is a tool that would address the following:

- Identify economic parameters that would add value and benefits to community.
- Evaluate the potential return on investment the program can handle at the scale of the site.
- Evaluate the community benefits that add value to both the site ecology and wider community.
- Evaluate the balance of energy system inputs and outputs. Determine if the project can be off grid or on grid.
- Evaluate how the site's natural resource use can be balanced and add ecosystem services value.
- Evaluate opportunities to reaching net zero. Ascertain the inputs and outputs the site is capable of handling.

Synthesizing Ecology, Culture, and Program into a Sustainable Design Concept

This part of the process is the most fun and the most inspiring, as synthesizing takes data and develops this information into a meaningful visual manifestation. It is simply the artful task of fusing the analysis data with the project vision into a physical shape, or, as what is known in design circles as "the big idea."

The big idea is the initial concept that begins to illustrate the vision taking it from abstract thoughts into a more fully realized physical possibility. Sometimes the idea is developed after the vision is created as a preliminary design concept and sometimes the idea is developed before thus becoming the thought seed for the vision. Just as music and lyrics often can be developed together or apart by the creative thought process, design can be created in the same way. The big idea becomes the germination seed of the project and acts as the form-giving framework for the project as the synthesis stage adds more layers of complexity to the design. The final design grows from the big idea over the course of the development process.

These are important elements in synthesizing ecology, culture, and economic factors:

- **Synthesizing culture**: The merging of ideation concepts with the project vision narrative provides for the marriage of vision with concept and form. The big idea rises from testing ideas of form for practical, aesthetic, and systemwide connectivity, interdependence, and innovation. When synthesizing culture, aspects of local and regional community benefits need to be considered. Providing connections to heal or enhance the food shed opportunities for the community play a factor in providing social value.
- **Synthesizing ecology:** Creating the framework plan from the site systems analysis of the green infrastructure systems and the ecosystem services provided. It includes watershed (drainage, water management and quality), soil health (nutrients, erosion control, carbon sequestration), energy (solar power and off the grid), waste management (zero waste, green waste, conversion of waste to food). The synthesis should evaluate how to apply these items from function into form, giving solutions.

ization of the vision or big idea into physical forms that respond to the synthesized data. In the analysis stage, the task is to collect the site and community's local and regional data, diagram the existing systems with an overlay of the proposed systems, and begin to define the initial program for the project based on the evaluation of the site's opportunities and constraints. Understanding the local context is key in this effort as, in order to connect to existing resource systems or infrastructure, it is important to know where those items are located and consider what it would mean to provide for those connections in the design. Ecology (site analysis), culture (program analysis), and economics (opportunities for value) are at the heart of this analysis stage.

Site Analysis (Ecology)

In creating a site systems analysis diagram of the proposed site, include the following:

- Begin with the climate data, site topography, and site survey.
- Identify and diagram the existing ecosystem components of watershed, soil shed, and vegetation biodiversity and habitat corridors.
- Identify and diagram the existing food system elements for the food shed.
- Diagram potential new connections to the existing natural systems as well as for any new systems or connections that are proposed.
- Conduct site systems analysis—green infrastructure: drainage, water management, soil nutrients and composting, potable vs. nonpotable water access, rainwater catchment, habitat and endangered species, etc.
- Explore off-grid versus on-grid for green infrastructure.

Program Analysis (Culture)

In creating a site programs analysis diagram of the proposed site, include the following:

- Identify and diagram the community social system connections that exist and are available.
- Form and function—begin by diagramming the components and their relationships to determine how they fit with vision and site opportunities. Evaluate how to tie to overall project if it is only a component of a much larger project program such as a new community.
- Provide scale diagrams of proposed project program to evaluate spatial impacts and connections.
- Analyze the urban ag opportunities for scale investment to select typology that fits the project site and vision.
- Evaluate the community benefits that might occur through the program elements.

Opportunities and Constraints Map for Integrated Decision Making (Economics)

Documenting the site and program analysis as an opportunities and constraints map is a way to merge the systems analysis diagram with the program analysis diagram in order to

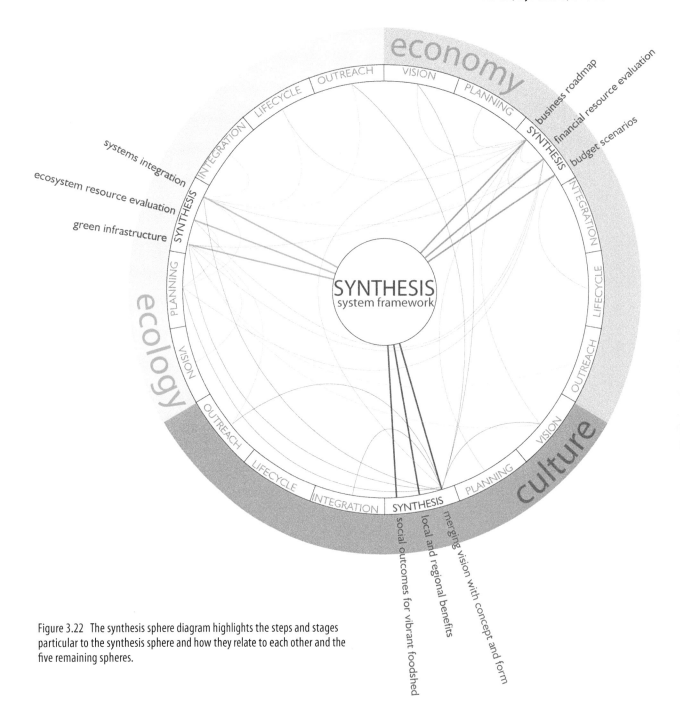

Figure 3.22 The synthesis sphere diagram highlights the steps and stages particular to the synthesis sphere and how they relate to each other and the five remaining spheres.

Systems Analysis—Evaluation of Opportunities and Constraints

The project idea has germinated; stakeholders have been identified; a project vision, mission statement, goals, and objectives have now been written; potential funding sources have been identified; and a site has been selected. Everyone is excited to move forward. The next step is to analyze and synthesize. This sphere begins with evaluating site and community systems resources and integrating the analysis into the refinement or conceptual-

that can tend toward looking messy if not planned in the design of the landscape. These items can be designed to be functional and aesthetic in the hands of a good designer. And in an urban ag landscape, while there are many urban farm components that functionally need to occur that typically are not thought of in aesthetic terms, that doesn't mean that they cannot be designed to be more aesthetically pleasing. Line, scale, color, form, taste, smell, and urban style can all play an important role in designing urban ag landscapes.

Gary Comer Youth Center Rooftop Garden, Chicago, Illinois

The Gary Comer Youth Center (Figure 3.23), in Chicago, Illinois, is an after-school learning center for the youth and senior residents of Chicago's south side. The project was funded by philanthropist Gary Comer, the founder of Land's End, who believed in giving back to the community where he grew up. Located in the city's economically depressed Grand Crossing neighborhood, the center provides a safe haven for education and recreation, while its garden provides a secure environment in which to experience the beauty and serenity of nature and feel safe from the violence below.

Figure 3.23 The symbiotic relationship of the garden with local restaurants enhances collaboration and education efforts.

Figure 3.24 The rooftop garden at the Gary Comer Youth Center, Chicago, Illinois.

The center's rooftop garden (Figure 3.24), by Hoerr Schaudt Landscape Architecture, is located on the third-floor courtyard over the gymnasium of the center's award-winning building by architect John Ronan. It serves as an outdoor classroom and laboratory, while introducing beauty and nature to enhance the educational environment. With an average soil depth of only 18 inches, the 8,160-square-foot garden produces in excess of 1,000 pounds of produce per year, which is put to use by students and in the center's cafeteria (Figure 3.25).

The garden is an aesthetic success as well as a functional one. With a dynamic geometric cadence that both responds to and enhances the building's aesthetic, the garden successfully marries landscape with the architecture (Figure 3.26). The edible growing space is divided into rows of varying size by circulation access strips formed from pavers of recycled milk carton material, each of which aligns with the third floor's window framing, uniting the exterior and interior.

Ornamental flowers enliven the space, and extruding circular skylights add accent and light to the gymnasium below. The vegetation has the additional benefit of reducing the building's climate control costs, while rooftop temperatures are still warm enough in winter to continue cultivation that would otherwise be impossible in Chicago's climate. An added benefit has been that because the garden is located over the heated gymnasium the heat helps to push the limit of the growing season. Because the garden is over the span of the gymnasium it also required more structural support for the load of the 18-inch depth of soil proposed for the agriculture. Peter Shaudt recalls the day that the structural numbers for this extra reinforcement came in at an additional $1 million to the budget and the edible garden was in danger of not being built. Because of Mr. Comer's vision for the center, however, it was deemed well worth the investment into the neighborhood. Mr. Comer's foundation also established an annual endowment for the center's operations.

Vision, Synthesis, and Form 119

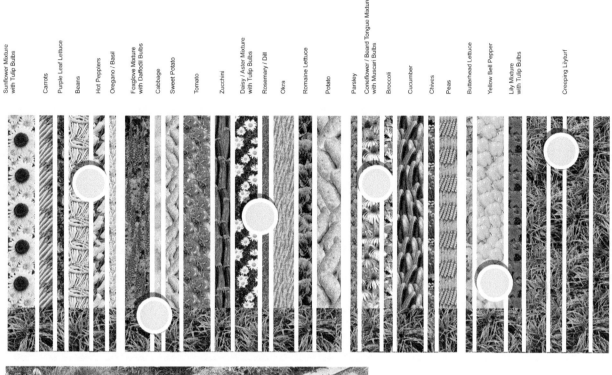

Figure 3.25 Illustrative plan demonstrating the poetic banding scheme for the Gary Comer Youth Center rooftop garden.

Figure 3.26 The planting bands are delineated by strips of recycled tire pavers and create a visually engaging space for users.

Figure 3.27 The afterschool programs provide an array of interactive educational opportunities for students.

The garden provides invaluable opportunities for the center's students to learn about horticulture, ecology, business, and environmental sustainability (Figure 3.27). The garden is visible and accessible from the third-floor corridor and classrooms. The center's garden manager Margie Hess has been instrumental in integrating the garden operations into the school's science and sustainability curriculums. Michelle Obama has visited the center and Chef Rick Bayless also has run a culinary program for the kids. Students also run Comer Rooftop Crops themselves, selling their produce to local restaurants while learning business and financial skills. Thus, the edible courtyard at the Gary Comer Youth Center not only enlivens the students' learning space but also raises their prospects for future employment and environmental leadership.

Design Team:

Client:	Comer Science and Education Foundation
Garden Manager:	Margie Hess
Architect:	John Ronan Architect
Landscape Architect:	Hoerr Schaudt Landscape Architects
Structural Engineer:	ARUP

Developing the Preliminary Urban Ag Business Plan and Business Model

It is important to develop a preliminary business plan early in the design process. It can be as simple as a draft outline at this phase of the process, but it does need to identify the key lifecycle considerations of the project, including the budget drivers—both incoming and outgoing, selecting the business model that best fits the project, outlining the future operations structure, the stakeholder decision making process, the future marketing and outreach potential, and the green job potential. Other items begin to get added to the business plan as it is developed throughout the process, however, without understanding the project's potential overall budget parameters, a project will not have a clear enough roadmap to tackle its own growth and development.

The task for determining the project's business model can begin as a lifecycle diagram to promote discussion and collaboration of the preliminary business plan with the stakeholders/advisory committee. The more that can be defined during this phase, the more the project's design will begin to reflect the business model choices desired and integrate the systems design accordingly. There will be a major difference between public and private developments when it comes to selecting the business model.

There are eight preliminary steps:

1. **Identify potential funding opportunities and constraints upfront—set up the outline for the funding system.** Funding can come from a combination of sources such as grants, fundraising, private–public partnerships, corporate benefits package, institutional donations, and more. Funding will be discussed in more detail in Chapter 6.

2. **Establish a preliminary budget framework.** How does the budget interface work? Let's say the design includes a compost bin as part of the waste stream system of the project. First, the cost for the compost bin needs to be identified in the business plan under the construction costs so it would need to be determined what type of compost bin is going to be used for the project to reflect that choice. There are a number of types of compost bins, ranging from simple to sophisticated options, that affect costs. In the discussion as to who will operate it and how will it be maintained, the business plan would include a cost for the operations and maintenance of the compost system as either part of paid personnel or a volunteer duty. Knowing the answer to these types of decisions as the project evolves in the design phase allows for the decision to cycle back into the project's design drawings and its final budget.

3. **Establish a decision-making process with stakeholders, the client, advisory board, committees, or subcommittees.** This is an important communications tool that will set the tone for the project. Communications models can evolve as the project evolves, but the key is that a communications protocol is in place.

4. **Determine the type of business model that should be discussed at the start of the project.** Even a number of potential options can be outlined upfront, with the goal to study the options at key decision points to refine the business model choices into one. The key is to clarify goals, objectives, and choices with the stake-

holders that best fit their value identification and vision. Questions to raise: Can you write a business model into the spec? Design the interface between, inputs, people, outputs, community needs, and cash? Does your business model include volunteers and a nonprofit, green jobs, grants, for-profit backyard farming, or a hybrid of community and for profit? What is the model that fits the needs of the community so that it will thrive into the future?

5. **Identify the lifecycle opportunities and constraints that best fit the stakeholders' value identification and project vision.** Evaluate this with the community's existing food shed opportunities and constraints. Question to raise include: Can the project heal a community that is currently a food dessert area? Can the project provide for food security? Does the project provide ecosystem services not being provided to the community?

6. **Identify the potential lifecycle operations structure.** Maintenance and management opportunities and constraints and understanding the expected lifecycle management of the system in this phase are all crucial to creating a design that functions properly. Questions to ask include: Who will be maintaining the landscape after the design team is done with constructing it, and how is it tied into the overall project operations? Will there be a paid position such as coordinator or kitchen staff, or will it be taken care of by volunteers or managed as a community-supported agriculture enterprise? What organization can your edible landscape support? Who are the community partners that can become part of the program development? What opportunities such as food banks or school lunch programs can factor into a zero waste management program?

7. **The preliminary production math for the return on investment, also known as the ROI, needs to be evaluated and designed into the lifecycle aspects of the operations and budget.** Questions to raise: What is the planned produce equation of the landscape? Hold a scale investment discussion—what size of landscape will provide what amount of yield? Testing the budget—can you increase ROI with tilapia aquaponics and greenhouses? How does the heat storage of water, protein, fertilizer, increased growing season, and vertical space affect the ROI? What size farm area will be needed? Farm sizes depend on the size of the "family"—typically, a 20-by-20-square foot area will feed a family of four. A 15-by-15-square foot area will feed a family of two to three. A 12-by-12-square foot plot will feed an individual or family of two.

8. **Identify the potential marketing, education, and outreach goals.** Have a discussion on how to incentivize and design a regenerative landscape that will benefit the community and local ecology. Have a discussion on green job training, education programs, community and business partnerships, school and institutional partnerships, and mentoring and longevity models.

Identifying the Potential Urban Ag Design Typology

Evaluation of the urban ag design typology that provides the right fit for a project is part of the collaborative conversations at the start of a project. Before the project vision moves forward into that stage for conceptualizing its physical form, the typology that is selected should be discussed with the stakeholders and core committee. It is valuable to discuss and evaluate the various types of typologies that may be appropriate for a project before making the final selection of what typology suits the project best. This discussion can be enhanced through the use of diagrams, definitions, or the use of case study imagery that helps to guide the ability to reach a decision. Until recently, urban ag landscapes have primarily been seen by the general public as urban farms or community gardens. However, the new wave of urban ag landscapes offers an expanded diversity of typologies as these landscapes continue to develop into new variations and hybrid forms.

There are many ways to break down the evaluation of typologies for urban agriculture. One way is through the evaluation of the participants (Figure 3.28) and how this exercise may uncover a potential strategy that could be utilized in the project's development phases. One of the first issues to understand is how identifying the project participants upfront helps to set up the communication systems framework as well as assisting in the providing of valuable information on the community's local and regional resources.

Figure 3.28 Identifying participants and stakeholders will give an understanding of the community's needs and resources and help dictate the appropriate design typology.

124 Designing Urban Agriculture

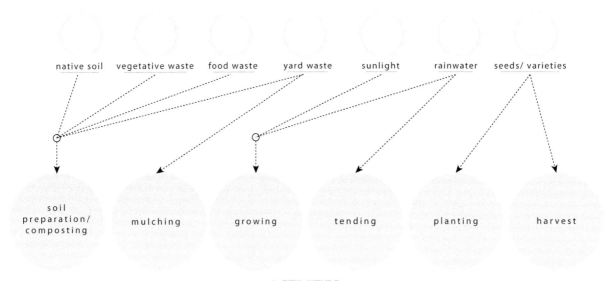

Figure 3.29 By identifying the resources and planned activities, a design hierarchy can be established and needed inputs named.

ORGANIZATIONAL CHARACTERISTICS

PARTICIPANTS:

- K-12 Students
- Parents
- College Students
- Homeowners
- Local Businesses
- Consumers
- General Public
- Under-privileged
- Employees

ORGANIZATIONS:

- Families
- Individuals
- Neighborhoods
- Schools
- Charities
- For-profit Producers
- Community Groups
- Restaurants
- Resorts

GOALS / OBJECTIVES:

- Personal Production
- Charity
- For-profit
- Fund-raising
- Education
- Research
- Environmental Mitigation
- Research

PHYSICAL CHARACTERISTICS

LOCATION:

- K-12 Schools
- Colleges
- Church/Temple/Mosque
- Community Centers
- Single-family Homes
- Apartments / Condos
- Neighborhood Plots
- Parks / Plazas
- Hotels
- Restaurants
- Urban Farms

PHYSICAL TYPE:

- Traditional
- Roof-Top
- Vertical
- Streetscape
- Greenhouse

SIZE/SCALE:

- Farm
- Garden
- Sub-Garden / Micro

OPERATIONAL CHARACTERISTICS

ACTIVITIES:

- Soil Preparation
- Planting
- Tending / Pruning / Weeding
- Harvesting
- Processing
- Distribution
- Sales
- Preparation / Cooking
- Planning
- Administration / Organzation
- Budgeting
- Donation
- Teaching
- Animal Husbandary

RESOURCES USED / PRODUCED:

- Labor
- Water
- Light
- Soil
- Compost
- Seed
- Produce
- Garden Waste

TERM

- Long-term
- Temporary
- Seasonal

Figure 3.30 It is important to recognize the defining physical, operational, and organizational features of the site to determine the appropriate design typology for the proposed system.

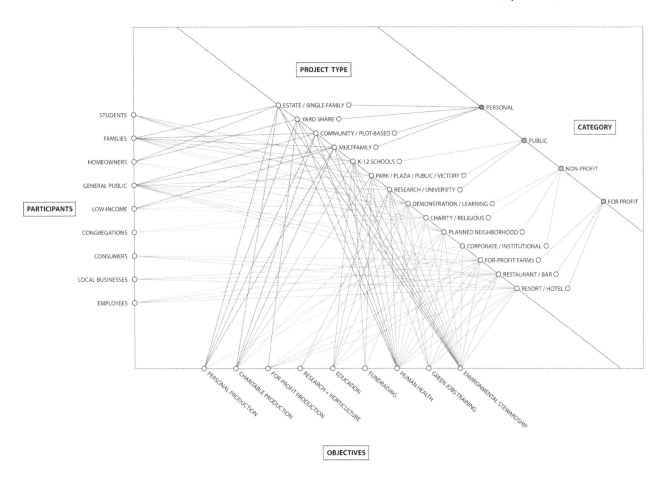

Figure 3.31 An appropriate design approach can be employed by identifying the connections between the physical and social characteristics of the site.

Another way of looking at typologies is by evaluating the number of connections between the resources and planned activities (Figure 3.29). This begins to map out system connections that exist and those that do not. It also can be used to determine which system connections need to be designed to support the framework and which connections are currently nonexistent but necessary to design in order to achieve the vision and goals to meet the system needs of the stakeholders and the greater community.

Typologies can also be further evaluated with additional complexity through their organizational, physical, and operational characteristics (Figure 3.30). Organizational characteristics include participants, organizations, and goals/objectives. Physical characteristics include location, physical type, and size/scale. Operational characteristics include activities, resources, and time. This type of evaluation begins to map additional layers of connections, interactions, synergies, and system relationships that begin to provide an integrated framework for the systems to engage with each other to build for more interdependence and more resiliency into the design outcomes. The more interdependence the system networks provide, the more resiliency is built into the system networks and the outcomes.

The urban ag typologies can also be further synthesized into a systems network organized by matching project type with the connections between the characteristics of participants, objectives, and categories that range from personal to for-profit uses (Figure 3.31).

The following typology descriptions are based on the range of project types that currently are redefining design within the urban agriculture landscape realm. This list is not conclusive, as the range of typologies including hybrids and variations is continuing to grow as innovation and knowledge of these landscapes expands. Another factor in the momentum for expansion of urban ag typologies is the propulsion for bringing more sustainable infrastructure systems into cities as cities begin to factor in climate change strategies and the role that ecosystem services provide in economic value to the city. Issues of human health, food security, and food deserts are also a major factor in changes ahead for urban agriculture to develop better food shed systems currently missing or broken in most cities.

One interesting characteristic that can be seen through all of these evaluation strategies is that there is much overlap and connectivity between the typologies. With future technological advances, we will also begin to see an expansion of the characteristics beyond what is currently being utilized today. For example, as green roofs and vertical walls come into more focus in cities of tomorrow, many of the typologies listed here will expand to newer models, variations, and hybrid typologies than currently envisioned:

- **Urban farms**—These are farms within the city or just on the edge of the city (peri-urban) that provide for income earning and food production, though some also provide recreation and relaxation. They are typically less than one acre in size due to zoning and policy legislation. Generally, the focus is on small-scale production of fruits, vegetables, and flowers for sale directly to consumers and restaurants or shared and bartered with a community organization's volunteers. The distinguishing feature is the diversity of crops on a small area of land. Some variables include bees and other small-scale animal husbandry aspects such as chickens. Generally, these farms are seen as very sustainable models since they promote energy-saving local food production and accessibility to local food in underserved areas of cities. Most urban farms are focused on organic food production, which allows for more intensive gardening on this smaller amount of land than a traditional rural farm. Some urban farms are also located on rooftops and use greenhouses and hoop houses to extend the regional growing seasons.

- **Community gardens**—Community gardens derive from when a diverse group of people in a neighborhood comes together to raise food. Each person or family has a designated plot. Most have agreements of use (and they find that natural outgrowths of the gardens are personal relationships, cross-cultural exchange, community development, beautification, environmental justice, crime prevention, leadership, and self-reliance for their neighborhood as a whole). One variation for neighborhood gardens is to be located on roofs. Some are driven by environmental justice concerns. Another newer variation is a private development version relating to high-density residential developments that incorporate a community garden space for their residents.

- **Research/experimental gardens**—These gardens are developed by a learning facility such as a university or research organization to test plants that might be suitable or more productive characteristics for the local environment and are open to the community to learn from.

- **Learning gardens**—Learning gardens promote food and garden-based education to communities by providing resources and facilitating partnerships for shared benefits. They can be school or community based. One example is the Learning Gardens Institute in Portland, Oregon, whose slogan is "Growing schools and communities together."

- **Demonstration landscapes**—A demonstration garden is a useful research, educational, and promotional tool for urban agriculture that generally has a specific purpose. They are demonstrating food shed issues, such as the large quantity of food one person can grow in a city backyard using intensive organic methods of cultivation, workshops, and hands-on experience to promote urban agriculture and urban gardening.

- **Edible school gardens**—School gardens are a variation of community gardens and are focused on combining education, stewardship, and nutrition. There are a number of approaches to edible school gardens, such as farm to table gardens and the Edible Schoolyard program founded by Alice Waters, which offers a culinary twist and a school lunch program addressing an underserved student population. In dense cities rooftops and vertical walls are offering schools more real estate to provide these urban ag landscapes. Some a connection with the school cafeteria or other community-based programs to support the roof infrastructure.

- **Food pantry gardens**—This takes many forms, but the main feature is the donation of fresh healthy food to local food pantries for families in need. Some are community gardens that coordinate the donation of excess produce to a nearby food pantry, others donate everything they grow or a portion of what they grow to a food pantry or community organization that then redistributes the produce to agencies.

- **Restaurant seed to table/farm to table landscapes**—These urban agricultural landscapes are as small as raised herbed beds outside a restaurant that is to be used by the restaurant chef in menu enhancement for food and cocktails such as Bar Agricole, in San Francisco, or as large as having a sizable portion of land adjacent to the restaurant and in other neighborhood locations that is grown solely for the use of the restaurant's menu. These restaurants' vision is in promoting locally sourced, healthy food that tastes good, is nutritious, and adds to the green economy of the neighborhood.

- **Edible hotel/resort landscapes**—Many hotels (Figure 3.32) are integrating food production into their landscapes to enrich the guest experience like the St. Regis Hotel by the SWA Group. (Figure 3.33). These can range from rooftop herb gardens to growing fruit, nuts, and vegetables for the restaurant, spa, or other amenity on site for a farm to table experience and may include beekeeping and wine making as part of the programming.

Figure 3.32 The St. Regis Hotel in Napa Valley, California, integrates the hotel and villas into the vineyard setting, incorporating the peri-urban agrarian landscape it into the guest experience.

- **Edible estates**—These landscapes are private edible gardens created by turning the front or rear yard, or both, into a productive garden space to supplement a family's needs for nourishment. Generally, a family will share or trade their produce with another family to increase the diversity of food choices. These landscapes are not usually grown for resale of the produce. A smaller variation would be edible balconies, where high-density city dwellers turn their balcony into an intensive urban ag garden.

- **Yardshare**—Yardshare is a variation of edible estates. Typically, it is a land-sharing or garden-sharing agreement between homeowners and an urban gardener or farmer enterprise or another neighbor for those families who want to grow their own food but lack space, time, skills, or ability to do so. It offers an opportunity to connect with neighbors to access unused yard and garden space. Thus, it might also represent community gardens built on private properties. The key distinction is that the owners are matched up with volunteers who tend a distinct plot in the garden. Homeowners get help while volunteers get access to land to grow their own veggies. There is an opportunity for this type of typology to expand through social networking sites and city networking sites.

- **Multifamily landscapes**—The urban ag landscapes organized around multifamily housing developments including apartments or condominiums can either be land based or roof garden based depending on the project location. Typically they are part of the open space land use criteria and seen as an amenity for the tenants.

- **Planned neighborhood food landscapes**—These urban ag landscapes have been designed to include a farmland component within the open space systems for a planned community. Organized around a vision of ecological stewardship and resource conservation, they also function as a social and economic benefit for a community promoting a mini food shed that can be connected to a regional food shed. Other potential benefits are food security and accessibility, promotion of a healthy lifestyle, community identity, and green job creation tied to resource management and conservation.

Figure 3.33 The St. Regis Hotel in Napa Valley, California, master plan.

130 Designing Urban Agriculture

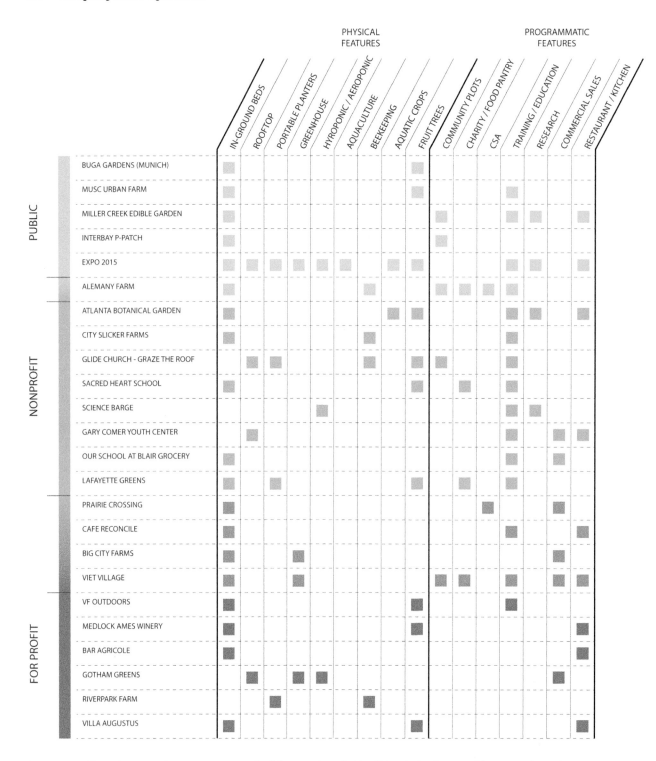

Figure 3.34 Different design typologies are appropriate for different projects. By identifying the key features of the proposed site, the organization can build a strong foundation and resiliency.

- **Company food landscapes**—These landscapes are being created as value-added amenities for employees use by companies that consider these food landscapes part of the company's sustainable vision. Generally, these company visions are focused on environmental stewardship, community service, social networking, and promotion of a healthy lifestyle.

- **Green infrastructure and streetscapes**—Urban agriculture as smart sidewalks, trails, paths, medians, and streetscapes offer a green component to the transportation infrastructure of a city and offer the ability to make streets safer, walkable, and pedestrian-oriented precincts, favoring foot traffic as opposed to automobiles. Streets as part of the city's circulation grid offer the opportunity to rethink the street in a greener light through urban ag landscapes. Transit corridors and greenways can be designed with ecological and cultural purposes and offer opportunities for multifunctionality and connectivity.

- **Parks and plazas**—All public open space developments such as parks, plazas, vacant lots, and leftover residual open space parcels offer opportunities for community-based urban agriculture, public–private partnerships, and festivals that focus on building community through food, and become integrated into the city's open space system.

- **Victory gardens**—A victory garden is a community garden in a civic space that includes vegetables, fruits, and herbs that are planted, maintained, and harvested by community volunteers. They represent "produce for the people by the people." They are a historical precedence for public urban ag landscapes reminiscent of medieval kitchen gardens, which have an even longer pedigree in history.

- **Wellness gardens**—Seniors and lifestyle-oriented communities offer an opportunity for using urban agriculture gardens for both fresh food and therapeutic benefit of the residents. These types of gardens can include edible planters that can be maintained by the residents for physical exercise and recreating with others, provide fragrance and visual stimulation to enhance memory, or provide fresh food for the cafeteria or restaurant where they gather communally.

The American urban-farm movement is seeing its biggest resurgence since its heyday during World War II, when victory gardens and kitchen gardens provided food for a rationed market, offered work, and educated children about agriculture. They're a shining example of what we as a country can grow in small spaces if pushed to the brink. These urban ag landscapes (Figure 3.34) and the remaining typologies represented in this book shed a light on the way to cultivating food system resiliency within our cities.

VICTORY GARDENS

History: Victory gardens were also called war gardens or food gardens for defense. They were originally vegetable, fruit, and herb gardens planted at private residences, schools, vacant lots, and public parks during World War I and World War II to reduce the pressure on the public food supply brought on by the war effort. In addition to indirectly aiding the war effort, these gardens were also considered a civil "morale booster" in that gardeners could feel empowered by their contribution of labor and rewarded by the produce grown. This made victory gardens a part of daily life on the home front.

Recent history: A grassroots campaign promoting the reestablishment of victory gardens has sprung up in the form of new victory gardens in public spaces, victory garden websites, and public blogs, as well as petitions to renew a national campaign for the victory garden. In March 2009, First Lady Michelle Obama planted an 1,100-square-foot (100 m^2) "kitchen garden" on the White House lawn, the first since Eleanor Roosevelt's, to raise awareness about healthy food and childhood obesity.

Design structure: Located at a city hall or other civic owned property (underused spaces in parks, streets, vacant lots, medians, parking lots).

Scale: Varies, but typically over 2,000 square feet.

Typical process: Forward-thinking local government officials and motivated volunteers.

Management structure: Community volunteers with a garden coordinator position typically filled by volunteer with master gardening credentials or other harvest-oriented expertise.

Benefit: Offers new opportunities for underserved populations and potential model that can help to reshape our food systems. They can promote healthy eating and bolster food security, and social innovation.

Resources

Holman, Peggy, Tom Devane, and Steven Cady. *The Change Handbook: The Definitive Resource on Today's Best Methods for Engaging Whole Systems.* San Francisco: Berrett-Koehler Publishers, Inc., 2007.

CHAPTER 4
Systems Integration and Connections

Medlock Ames Wine Tasting Room, Healdsburg, California

The Medlock Ames Tasting Room, in the Alexander Valley near Healdsburg, California, serves as the face and storefront for the Medlock Ames Winery. The winery, founded by friends Chris Medlock James, the winery visionary, and Ames Morison, the visionary winemaker, is known for its sustainable production of small-batch wines. Its nearby all-organic, solar-powered 375-acre ranch and vineyard produces vegetables and olives along with grapes. Likewise, the tasting room serves as a showcase not just for wines, but for sustainable construction, the farm to table agricultural approach, and design that is both beautiful and functional.

For years, the property was home to the Alexander Valley Store & Bar. The owners of Medlock Ames have built on this legacy not only by constructing their tasting room and separate saloon in the footprints of the former store and bar, but by using the materials of those abandoned buildings in their new facility. The site makes use of repurposed fencing (Figure 4.1), reclaimed wood, and recycled concrete, enhancing the patina of the landscape and referencing the working nature of the site. Located within the peri-urban outskirts of the town of Healdsburg with its contextual setting of vineyards (Figure 4.2), the modernist lines of the landscape are rooted in the agricultural vernacular.

Figure 4.1 The lines of Medlock Ames invite the visitor to explore the food gardens and beyond.

Figure 4.2 The water-conserving plant palette is of the vernacular of the wine country region.

Figure 4.3 The eighteen 9-foot-square galvanized steel raised beds are the heart of the edible garden. Herbs are paired with light-tasting food bites and served in their specialty cocktails at the bar.

Figure 4.4 A diagram of the stormwater management strategies of the landscape.

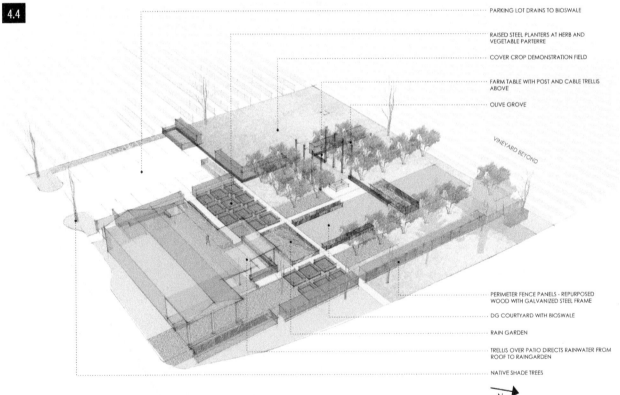

Much like the tasting room's interior, the three-quarter-acre modern landscape outside is bold, simple, and clean. A ground plane of decomposed granite is punctuated by paths of linear concrete pads, while low board-form concrete walls carry the architectural language of the tasting room's patio into the garden. The space features an organic vegetable garden of eighteen 9-foot-square galvanized steel raised beds (Figure 4.3). The ground plane of the garden slopes, so the beds were set below grade and leveling brackets used to make sure they align with each other. The landscape also includes an open courtyard and olive grove with a pizza oven and family-style table to host events throughout the year or used by the daily wine country picnickers visiting the tasting facility (Figure 4.4).

Ecology (Figure 4.5) plays an important role on the tasting facility property. In order to manage stormwater on-site, the garden drains to a rain garden, while the nearby parking lot drains to a bioswale. The landscape infrastructure also includes the collection of rainwater from the roofs for use in the landscape and percolation into the Russian River watershed. The river is about a quarter-mile away. Elements from the infrastructure were dually engaged as an aesthetic feature to visually display and expose the process of the systems. A trellis at the building guides water overhead and down chains to the permeable surface below (Figure 4.6). The plant palette features drought-tolerant native shrubs and grasses, along with a native fescue lawn meadow, that recall the natural vernacular of Alexander Valley, providing a real sense of place to the locale (Figure 4.7). The site's soil was enriched with all-organic amendments and compost that is created on-site.

Figure 4.5 The outdoor extension of the tasting facility extends into the food garden.

Figure 4.6 Rainwater is collected from the roof and flows into the extensive rain garden feature outside of the dining area.

Figure 4.7 The olive orchard frames the edge of the garden and offers sanctuary during the hot summers.

The garden exemplifies the greater commitment to responsible land stewardship and wildland preservation held by Medlock Ames; the one-acre site is but a fraction of the 375-acre property, of which 75 percent is managed woodlands, grasslands, and chaparral ecosystems. The remaining 25 percent is employed for organic farming of grapes, olives, and vegetables. The garden in the one-acre tasting room site is planted with seasonal crops to celebrate the annual agrarian calendar and provide the kitchen with the seasonally appropriate produce. Integrated into the edible garden is a variety of plants chosen for their draw to pollinators, a key necessity for any producing landscape. These include Queen Anne's lace, mustard, buckwheat, and sunflowers. Many herbs known for deterring pests were also planted—not only for their culinary use but to minimize any need for applications of pest control. This is especially important since the garden is open to the surrounding open space system and accessible to people and wildlife alike.

The garden also features a communal dining space (Figure 4.8) with a custom reclaimed fir table and benches, and an outdoor pizza oven. String lights run overhead, suspended by large cedar poles. An olive grove of 24 trees alludes to the larger olive orchard at the Medlock Ames ranch, from which they produce as much as three tons of olive oil in a season. Wine club members are invited annually to harvest the olives from the orchard and shown how to brine them. They even get to take their own batch of olives home. This type of interactive engagement with food from fresh organic ingredients is part of the environmental stewardship philosophy to gently teach while having an enjoyable time.

Figure 4.8 The outdoor garden area is built from repurposed materials.

Figure 4.9 The indoor–outdoor relationship of the architecture and landscape is paramount to the California wine-tasting experience.

Produce from the garden is used for pizza toppings, salads, and other culinary treats, and is paired with wines in the tasting room (Figure 4.9). The bartenders use citrus and herbs to make their custom cocktails. The tasting room manager also jars produce from the garden, including beets, beans, and tomatoes, which are sold in the tasting room. Employees gather together once a week to harvest the produce at the ranch's farm garden and get to take home whatever they'd like to enjoy with their family. The remaining produce not used in the tasting room, bar, or for weekend gatherings is sold weekly at the local farmers market stand outside of the tasting room.

Design Team:

Landscape Architects:	Nelson Byrd Woltz Landscape Architects
Local Design Consultant:	Thomas Woltz and Alexis Woods
General Contractor:	Earthtone Construction
Landscape Contractor:	Creative Environments

PART 1
SYSTEMS INTEGRATION

An Integrated Systems Approach Continues Throughout the Design Process

Once the sustainable framework plans synthesizing the design ideas and preliminary system information have been established, an integrated systems approach as diagrammed in the systems integration sphere (Figure 4.10) continues throughout the project development. Integration is a holistic term that implies achieving harmony, blending, or fusing elements into a whole. It is based on Aristotle's idea that the whole is greater than the sum of its parts. The Greek word *synergy* means working together. When systems work in synergy with each other they produce a result that is not attainable independently. Planning for system connections will not always provide a predictable outcome. In the case of ecosystem function, the systems are moving toward a more organic model of open-endedness that allows for flexibility and adaptation and away from the traditional model of precise stability and control. This means that current scientific thought sees ecosystems as open systems that are self-organizing and thus unpredictable at times. Change is part of living systems and an acceptable state of dynamic flow; uncertainty is part of how the system performs.

> With urban agriculture landscapes, the ultimate sustainability goal is to design systems that allow for accommodating a dynamic of interdependence or, more specifically, a landscape that attains a regenerative level of performance.

Every design decision and adjustment has the potential to make the systems and their networks function or not function on a more integrative level. As previously discussed, interconnectivity and interdependence are two of the primary ways that systems work together or within themselves. Fluidness, flexibility, and cyclic responsiveness are the key words for guiding this portion of the design process as the project design evolves into its constructed form.

To identify potential system connections, refine system diagrams begun during the creation of the preliminary design framework plans and develop an urban ag systems matrix to identify the potential system connections. These tools can be instrumental in the mapping and tracking of the proposed system connections. They are also useful in identifying the places where the connections may not happen as planned. Through the information gleaned from these tools, the design can continue to be adjusted and modified as it moves ahead.

Systems Integration and Connections 139

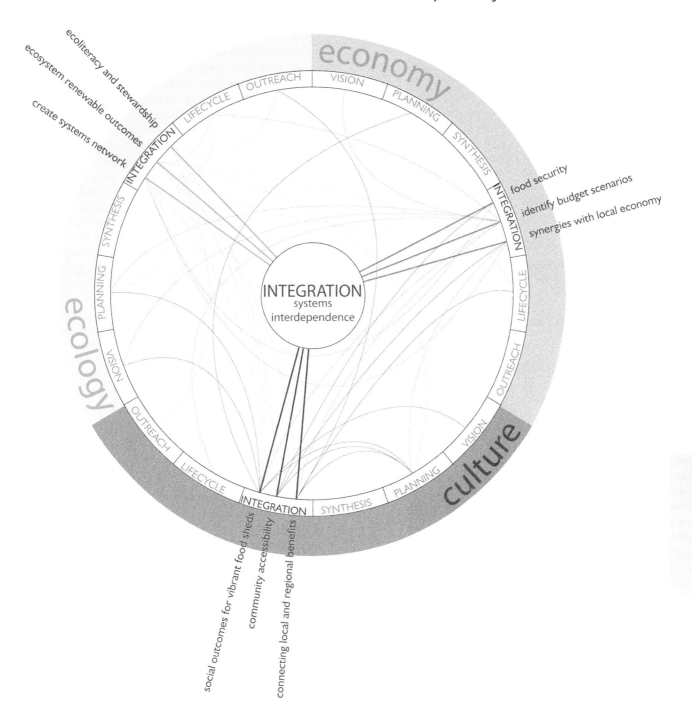

Figure 4.10 The diagram for the systems integration sphere.

> **ECO-BENEFITS OF AN INTEGRATED SYSTEM APPROACH FOCUSING ON ECOLOGIES**
>
> - Cleaner air
> - Reduction in pollutants
> - Increased habitat networks
> - Improved water quality
> - Watershed protection
> - Improved water infiltration
> - Water conservation
> - Increased soil health
> - Reduction in erosion
> - Increased biodiversity
> - Carbon sequestration
> - Reduction in heat island effect
> - Protection of renewable resources
> - Ecosystem resilience
> - Reduction in waste
> - Increased energy capacity

Mapping the System Connections

We have already discussed that an urban ag landscape is most successful when based on an integrated system approach that links natural systems with built systems to achieve a food-producing landscape that benefits the community. With an integrated systems approach, the evolution of how the systems are designed for integration into the final design output is what supports a living system. How the systems are refined into the final design becomes more achievable if the proposed system connections can be mapped. The urban ag resource systems matrix (Figure 4.11) is a mapping tool that can be used to help evaluate the potential connectivity aspects of the proposed systems. The more connections that can be achieved, and then later maintained, the more the systems network will function with a higher sustainability outcome.

This matrix is organized into three core sustainable building block categories: ecological, cultural, and economic. Most discussions on sustainable systems are generally based on these three categories to demonstrate that balance between the three categories equals achieving sustainable harmony. We have used these three categories as the basis for our urban ag process spheres described in Chapter 3. Some projects will have more system parts to integrate than others but all fundamentally rely on these sustainable system building blocks at their core. Before a project begins to integrate and link the systems within its specific scope, a working understanding of what is considered in each system is necessary so that the appropriate design strategies for integration can be more readily determined and harnessed.

Systems Integration and Connections 141

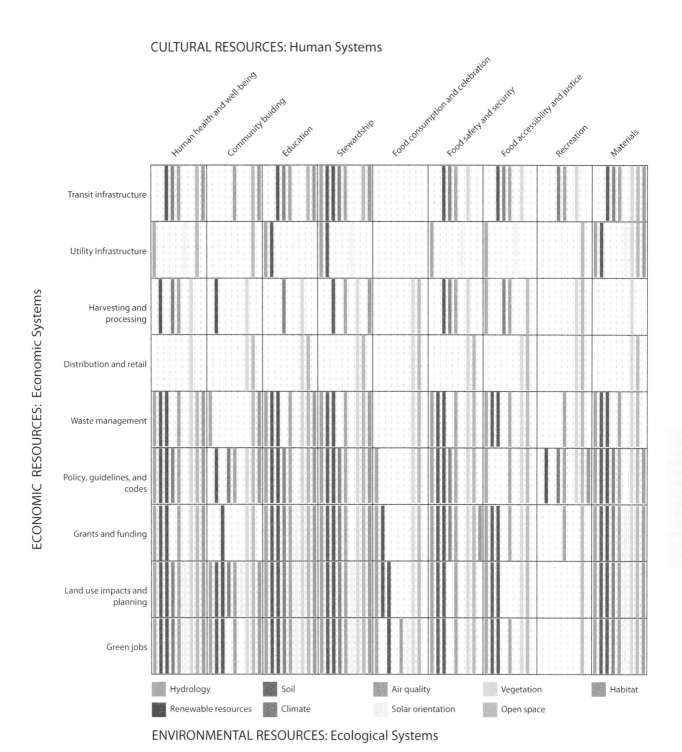

Figure 4.11 The urban ag resource system matrix.

Environmental Resources

The urban ag resource systems matrix in Figure 4.11 includes the following environmental resources.

Ecological systems:
- Hydrology—water input and output
- Soil—microbiology and nutrients
- Vegetation—food production and biodiversity
- Habitat—relationships of beneficial and species dynamics
- Climate—temperature and moisture
- Air quality—ability to improve/location
- Solar orientation—location and seasonal
- Renewable resources—local, natural sustainable materials used wisely
- Open space—connections between human health and ecological health

Cultural Resources

The urban ag resource systems matrix in Figure 4.11 includes the following cultural resources.

Human systems:
- Human health and well-being—physical, biological, and mental
- Community building—connectivity and vitality
- Education—relationship between health and nourishment
- Stewardship—fostering connections between health and nature
- Food consumption and celebration—nourishment of body and soul
- Food safety and security—connections for protecting human health
- Food accessibility and justice—nourishment for all equals healthier life/healthier communities
- Recreation—active and passive relationship on human health
- Materials—local, manmade, sustainable and nontoxic have positive impacts on human health

Economic Resources

The urban ag resource systems matrix in Figure 4.11 includes the following economic resources.

Economic systems:
- Transit infrastructure—accessibility to food and jobs
- Utility infrastructure—integration of green infrastructure
- Harvesting and processing—connections between consumption, food accessibility, and green jobs
- Distribution and retail—connections between consumption, food accessibility, and green jobs
- Waste management—connections between compost, recycling, and green waste as an input and output (with distribution to food banks and school meals)

- Food and health policies, guidelines, and codes—fostering and building healthier communities and food security
- Grants and funding—relationships between community support, vision, and economics
- Land use impacts and planning—fostering and building resilient communities
- Green jobs—connections between education and prosperity

Monitoring the design modifications at various points of the design process in the matrix format helps to determine if a connection is going to be possible to achieve or not achieve. There will also be a number of connections that may be hard to predict until the project is completed and operational. There will be others that will remain intangible since there are not yet metric tools available to measure dynamic relationships. Some system connections may not be known at this point in the process. Perhaps the operational options are still a group of alternative scenarios or perhaps available funding grants or partner organizations have not yet been identified. By monitoring the systems throughout the process, these known and unknown facts become a valuable part of the input–output testing process for urban agriculture projects, thus allowing for continued collaborative conversation to successfully resolve them.

Site Strategies to Increase System Connections

As the design is developed, site strategies to increase system connections should be studied to determine which are going to be the most appropriate for the project's specific site and location. Testing the strategies as the project progresses provides the framework for a fluid, dynamic systems design. Sometimes it may be hard to predict the exact outcome, but the systems matrix remains a useful tool to chart them. In many cases, parts of a project may be phased over an extended time frame so charting the overall connections and the interim connections also provides a roadmap for the future as the project raises the capital funds to realize its goals.

Examples of site strategies that can be incorporated into the design to promote more system connectivity or synergies into an urban ag landscape include:

- Providing habitat for attraction of beneficial insects for integrated pest management and plant health
- Stormwater harvesting for agricultural irrigation
- Stormwater quality treatment to increase health of local watershed and increase low-impact development solutions on site
- Establishing habitat corridor connections within an open space network of the community that also connects pedestrians to the project site
- Increasing soil nutrient health to promote increased vitality and nutrition to organically feed productive landscapes, combat urban erosion, and sequester carbon
- Increasing the biodiversity of the productive plant palette to add biodiversity to the surrounding community
- Using both vegetation and soil solutions to address carbon sequestration and improved air quality for neighborhood
- Using vegetation and structures for addressing micro climate control and vertical growing solutions to maximize spatial availability
- Locating productive landscapes in food desert areas to address food justice, food accessibility, and community health

- Designing urban ag landscapes to connect people to nature within the city to address human health and well being
- Providing nutritional education and nutrition-rich produce to address local human health issues
- Designing the urban ag landscape to create local green jobs through volunteer training, education and business development
- Designing for capturing renewable energy as an output of urban farm operations
- Providing for local sourced materials to increase local economic benefits
- Designing for on-site lifecycle waste management to increase environmental and economic benefits to local community and reduce strain on city infrastructure
- Designing an urban farm that supports and connects a community to both food and open space

Our School at Blair Grocery, New Orleans, Louisiana

Our School at Blair Grocery is an urban farm and education center located in the Lower Ninth Ward neighborhood of New Orleans, Louisiana. The school was founded by Nat Turner, a schoolteacher from New York City. In response to the devastation by Hurricane Katrina, Turner began taking students on volunteer relief trips to the Lower Ninth Ward in 2006. During his trips, he noticed a significant unmet need for healthy, fresh food in the neighborhood, and decided to stay for good in 2008 with the goal to tackle this food desert problem.

Turner found a home for his new endeavor when he signed a 10-year lease on an abandoned former grocery

Figure 4.12 Education at Our School at Blair Grocery focuses on participation with neighborhood youth.

store (Figure 4.12)—an appropriate place, given that the downfall of such locally owned stores is one reason the neighborhood became a food desert. Though Turner was a novice to farming at the time, four years later Our School at Blair Grocery has grown into a significant, multifaceted operation. The site hosts hoop houses, a large composting area, and twelve 4,000-square-foot plots, growing tomatoes, arugula, mirliton—also known as chayote—beans, peppers, hot okra, and more. The school has even expanded to land on an off-site ranch where it grows large quantities of field peas for sale to local markets.

In an effort to tackle issues of joblessness and illiteracy among the neighborhood's youth, Our School at Blair Grocery is as much a school that teaches ecoliteracy as it is a farm (Figure 4.13). Local youth become students as well as farmers (Figure 4.14), and have mastered marketing and sales in

Figure 4.13 Composting is part of the sustainable farm practices for building healthy soil.

Figure 4.14 Self-discovery is just as important as classes.

addition to farming. The students market and sell their produce to markets and restaurants around New Orleans every week, even negotiating prices themselves. This approach allows the students to take ownership of projects. For example, 15-year-old Sabrina coordinates all of the school's farmer's market activities, and 16-year-old Sam has taken charge of their fledgling mushroom business. The whole community becomes engaged stakeholders who are vested in the health and well-being of all (Figure 4.15).

Figure 4.15 Community members help to build new planters.

Local Codes and Policies

Research of local codes, guidelines, and policies that affect urban agriculture decision making should continue to be a key step in this part of the design process. Urban agriculture policies are continuing to be adapted or incorporated into the policy framework of cities and communities as the demand for these landscapes continues to grow. In places where there are outdated codes and policies, a project will not be able to go forward unless the existing policies are updated or modified. Knowing this information upfront is key to knowing the steps to successfully getting a project off the ground. Policy and advocacy issues will be discussed in more detail in Chapter 6.

Research in the design development process should include discovering if there are new types of technologies available for incorporation into the project, and if they are permitted by your local codes. In the field of urban agriculture, new products and technologies crop up in the marketplace at a very fast pace, especially in the "DIY," or do-it-yourself category, where social media can add to the instant promotion of the latest urban ingenuity. For example, on the TED website, a DIY vertical window project in New York is a fresh, ingenious way to grow food in a small apartment with only a window required and a few simple materials. In a short time, you could have an edible landscape in your very own home.

Using Feedback Loops to Keep Project on Track

As a project develops, stakeholders and team members should meet to build a feedback loop into the communication process to help them fine-tune the design for form, function, and performance. The project's business plan model is refined in this phase and documented with changes and updates as decisions reach the next level of decision making on the lifecycle aspects of the project. If a business model plan has not been developed by this phase, it is important to write a draft for the client and/or stakeholder input before moving forward. Depending on the size of the project, and if it is a private development, it is possible to approach the model as a draft plan with a number of business operation and funding options that can be explored as the project moves ahead. Each of these elements of the design process increases the likelihood of designing a more unified and holistic systems network that addresses the project on multiple levels.

Systems integration can be further enhanced for a project if a sustainable rating system such as the Sustainable Sites Initiative, SITES, LEED™, One Planet Living, or Biomimicry is being pursued. As rating system credit exhibits are developed, the design components can be further documented, enhanced, and discussed by the team, promoting further multidisciplinary connections and linkages that might not have been thought possible. The more inputs to evaluate at this stage in the process, the more opportunity there is for creative innovation to occur. This allows for the potential to increase a project's connectivity strategies as the sustainable strategies within the design become more finalized through the construction documentation and implementation stages.

Systems coordination with multidisciplinary design teams and the project stakeholder groups such as neighborhood and community groups, nonprofits, school districts, city departments, or even for-profit enterprises will vary in terms of the organization, but all must begin with setting up a clear channel of communication. It is important for everyone involved to feel informed and feel they are part of the process. Building a simple feedback loop into the communication process is important for identifying and verifying the connections from within the project (inputs) into the community or citywide systems (outputs). The lifecycle for communications and monitoring of the systems coordination must allow for eventual evolution from the design phase, through the construction phase, and through its ongoing operations after the designer is long gone. Setting up clear communications will help to make the transition from the design team leading the process to the client or stakeholders' management team leading the project's ongoing lifecycle operations and programs.

Communication inputs and outputs through the design development process include the following:

- Plan for collaborative conversations (Figure 4.16) and feedback loops throughout the process to keep stakeholders informed and engaged in the outcomes.
- Review the mission, goals, and objectives at various phases to test that the vision is being met.
- Review the design evolution, aesthetics, and metrics along with the opportunities and constraints to ensure design solutions are best suited to meet the project vision.

Figure 4.16 Collaborative conversations can be documented by sticky notes and other means to keep the dialogue moving and give the stakeholders a part in the decision-making process.

- Review the system connections at various checkpoints in the process to test for system integration.
- Develop and monitor the urban ag business plan throughout the process to test that the budget goals are being met and vetted with stakeholders.
- Do the production math at various stages of the design process to determine and test the return on investment (ROI) and to provide a format to continue the evaluation process annually once landscape is operational.
- Develop the schedule and milestones for the necessary drawings or documents to meet the local approvals, permits, bidding, installation, and review process with the design team, city departments, and stakeholders.
- Determine the components to be included for the lifecycle process of the project in order for the stakeholders to understand what the lifecycle operations expectation is and to allow for the design to integrate them early on in the design process.
- Determine who will be providing the preliminary seasonal guidelines for creating the harvest plans appropriate to the location and integrating those expectations into the design plan.
- Plan upfront for the harvesting of the landscape and rotating of crops into the design framework and understand how the landscape will be maintained and operated before it is constructed.
- Discuss the seasonality lifecycle plans to build future flexibility into the design and considerations of the climate and users.

Refining the Urban Ag Business Plan to Address Lifecycle Performance

It is important during the design phase to revisit the urban ag business plan, the project's roadmap to success, periodically with the client or stakeholder group as the project develops. The goal is to have the final draft of the plan handed over and transitioned to the client or stakeholder group, who will be in charge of overseeing the landscape once it is installed. One way to approach the business plan during the design and planning phases is to provide the business plan as part of the design services for the urban ag landscape component. If the client or stakeholder group does not yet have answers to these questions, begin by asking the question: What type of organization framework can the food landscape support? What is the model that best fits the project so that it will be successful, vital, and thrive in the future?

The business plan is primarily a vision and budget outline that presents an operating framework for the project that reflects the mission and values of the stakeholders and the expectations for its functioning as an eventual business enterprise. Designing the business plan to address the expected or proposed interface between inputs, outputs, client needs, community needs, cash flow expectations, annual funding and grant goals, benefits, volunteers, staff, green jobs, education and outreach, deciding if the project will be nonprofit enterprise or a for-profit hybrid—all of these elements of the business plan need to be determined as early as possible since they do affect design choices that should get incorporated into the final plans.

Another approach in providing this effort as a service is to consider the business plan as part of the project's evolution into lifecycle performance specifications. Is it possible to clarify the original vision, goals, and objectives to determine if they still reflect the client and stakeholder group's values by the end of the project documentation and installation stages?

The exchange of ideas and understanding about the expected lifecycle operations and management of the design and its systems is crucial to creating a design that functions properly and meets the project's vision.

In the discussion on the operations and maintenance of the urban ag landscape, the business plan should include a preliminary budget for the ongoing operations and maintenance, starting with year one and then estimating at least year two and three. This provides a baseline that can be used for determining if the vision goals will be able to be met year to year or what the time frame will be for meeting them. For example, in evaluating items such as how the compost system upkeep will be accomplished—would this be performed by paid personnel or be part of the landscapes volunteer workforce? Will the edible garden have a paid facilities manager, an educator to develop the education programs? Knowing the answer to these types of questions as the project evolves in the design phase allows for the decisions to cycle back into the project's design drawings and be included in its final design *and* budget.

Other questions to discuss include, but are not limited to, the following:
- Who will be maintaining this landscape, and how is this landscape tied to the rest of the project if the urban ag landscape is only one part of a larger project, such as a corporate campus?
- Will the urban ag landscape be a nonprofit entity or a for-profit entity?
- What is the annual operating budget, and where will the funds come from?

- Will a grant writer need to be hired?
- Will the food landscape be maintained by volunteers or paid staff? If volunteers, who will be in charge of that workforce, and how will they be organized and communicated to?
- Will there be a paid position for a garden coordinator, urban farmer, or kitchen staff?
- Will the landscape be part of an existing neighborhood community-supported agriculture (CSA)?
- Will it be tied to and managed by an adjacent restaurant or learning institution?
- How will excess harvest produce be handled? Will that be tied to a local food bank or a school lunch program, or will it be shared with volunteers?
- Can the landscape be combined with other land parcels as one entity for scale aggregation?
- Will there be an on-site market stand, or will produce be sold at a local farmers market?

Preparing the Preliminary Budget

Preparing a preliminary annual budget is best begun with a description of the elements for both construction and anticipated operations that are important for developing a preliminary estimate. Budgets will depend on the size of the operation and the business goals that were set in the vision phase and will depend on the scale of the project.

Preliminary budget breakdowns typically include the projected:
- Estimate of construction costs of raw materials
- Estimate of projected costs of labor
- Projected total for capital costs
- Phasing options
- Potential for volunteer labor and in-kind services
- Projected costs for managing landscape and the budget accounting
- Projected costs for operating and maintaining it—daily, weekly, monthly; Annually: first year and beyond for at least three years
- Projected costs for educational and training programs
- Estimate of costs for tools and storage
- Estimate of costs for seeds and new plants on a seasonal basis
- Projected costs for waste management
- Projected costs for marketing and branding

How the budget interface with the design works is explained in more detail in Chapter 3. The important criteria to follow is that every decision is connected to the inputs and outputs; so, for example, a decision made on how to collect and recycle waste also can impact the projected operations and management costs depending on the solution. Working with a contractor who can validate or provide probable costs is a real plus, especially in the early stages of a project, to set ballpark targets. Understanding how the urban ag landscape will most likely be operated is also a great asset in order to set the preliminary budget.

Testing the Return on Investment

The return on investment, or ROI, is an important concept to understand. It provides a means to test the budget options and address the basic question of how much can I produce in a certain amount of space—what size farm or food landscape do I need, and how much will it produce? Elements that may affect the production numbers include the heat storage of water, use of protein amendments, type of fertilizer, incorporation of elements to increase the growing season such as use of tilapia or greenhouses, or increased square footage of growing area by using vertical space, not just horizontal space. The size of the productive landscape area, how intense it will be planted, and what is going to be planted, all factor into the amount of produce that can be harvested and the size of family or community that can be supported by that landscape. When calculating the potential ROI of the urban farm's production, it is important to assess it holistically and site specifically. When you are identifying the local ecosystem assets of an urban landscape, it is important to include the local knowledge base of people who are already gardening and farming in the area. The experiences of these people are invaluable to you in assessing an accurate ROI potential for your urban farm's production. Talk to people in the neighborhood and find out who the most experienced gardeners are and then interview them. Find out what crops they grow and how much they produce in the square footage they have. Find out what style of gardening they do. Is it biointensive square foot gardening, permaculture, no-till, organic, irrigated? Find out all the logistical details that make up their system of food growth and maintenance, the size of the gardens, the type of food they grow, how much labor they put into it year round, and the poundage of food that is produced. It is also important to find out the resource inputs that they put into the success of the growth. Calculate the labor, the compost, mulch, water, insect controls, seeds, and all other inputs that cost time and money to use. Of course, it is also essential to find out the outputs that they are achieving with the inputs. Find out how much of each food is produced, how much material is needed for composting and mulching on site, what type of community outreach and marketing is achieved, which crops have the best poundage per square foot, and what sells for the most per pound. This will start to give you the information needed to calculate potential ROI into the design as a key factor to determine the success of your design. Since each urban farmer does things differently, it is important to interview as many folks as possible to identify a large list of diverse possibilities to give yourself design options for your site. This will give you a good list of options when designing your new system for your site specific and stakeholder needs.

The data from the farmer interviews can then be applied to the goals of the project that you may have for the business plan. You can start to decide what options will work for your gardens based on how much square footage you have, how many inputs the site can afford on a yearly basis, what maintenance is possible, and what production is possible based on what you choose. The design will also be dictated by the way the site will be managed by owners, the community, employees, volunteers, and so on. Consider the talent level of the people who will be working on the garden, and design in a style of farming that matches or is trainable to those workers. Adding these factors together will help you create a holistic balance sheet of all the factors that determine the inputs and outputs of the garden while including the human skill and budgetary realities.

Always go back to the people you interviewed and the people who will be managing your design to ask them if what you are calculating is likely to succeed. The design you put together will likely be different from their site, and it is important to talk through how things will grow, given the way you have designed it. Each design will have different dynamics within it when putting the different design elements together. Talk through these dynamics with the local farmers to see if they have advice or identify any red flags.

They will be able to tell you what they think will work, what they are willing to manage, what is likely to not work, what is too idealistic, and what is possible that you may have missed for greater opportunities. Don't think you know how to farm because you have done your interviews—ask for help from the farmers all the time to ensure you are rooted in the reality of what it takes to be successful. It may be important to include a few farmers on your design team that are very good at the elements of farming you think you will implement in your design. One biointensive farmer, one good market farmer with high production success, one that has done well with engaging communities and using volunteers might be some good examples of a diverse team to help you. Adding their perspectives can help you think through the complexities of the holistic ecological farm design and set it up for success by including the planning and maintenance in the design along with the beauty and functionality of the property.

It is important to match the new farmers with the right design that matches their abilities and needs production in the space provided. The design can set the owners up for success or failure. It can make the production and ROI easy or hard. The site should not need too many inputs from off site to be imported in order to ensure the levels of production needed to hit the profit goals.

Selecting the Appropriate Installation Methods

Designing and implementing a sustainable food system takes contractors who can do more than just read plans and implement the specifications. They need to be able to understand what sustainable methods are for the engagement with your stakeholders, the community, the soil installation, and all the little decisions that need to be made throughout the process. The site-specific conditions of your project will need to be assessed continuously through an understanding of the complex set of goals a sustainable food system can have in the short term and long term.

They don't have to know everything that your design team and stakeholder group knows, but they do need to be able to have enough understanding of stakeholder management, what makes a sustainable landscape, how to assess and regenerate soil systems, and collaborative communication in order to be an effective partner on your team, not just the installer. Contractors have potentially many touch points with the community and client stakeholders. They need to understand they are a key part of the success of the project for their expertise and engagement with the community. They can really be a bridge between the design and the ongoing care of the project. Most importantly, however, is the ability for the contractor to be fully honest, transparent, and trusting in order to navigate all the considerations that might go into a project. An example is how the selection of the specimen olive tree that anchors the VF Edible Garden started with a visit to the Olive tree farm to determine the best tree for the site and then the farmer's advice on the transplanting process with the landscape contractors (Figure 4.17 and Figure 4.18).

Figure 4.17 At VF Outdoors in Alameda, California, a heritage California Mission olive tree being placed in the field before the remaining edible garden is constructed. Scheduling is an important part of the process.

Figure 4.18 The tree is in place before most of the employees arrive, though early birds added to the excitement as the tree location was finalized.

This might seem like too many needs from a simple contractor who is just going to install the system. This would be true if the contractor just had to install pipe, soil, plants, mulch, and irrigation controllers. Installing a sustainable food system requires more knowledge and a long-term partnership that is brought into the project to stick with it as it evolves. It is important to choose a contractor to be involved near the beginning of the design process to give input into ongoing discussions of maintenance, budgets, training, skills needed, and creating systems for the community to implement when the design and construction team is gone. You want an implementation and thought partner that is bought into the project long term to help the project succeed. You will be building dynamic systems; it is not a static landscape that will give pleasure to the community for 10 to 25 years and then be redone again. You are looking for a long-term partner to build food cities that are dynamic ecosystems.

When asking for proposals, ask contractors to describe their understanding of ecological systems, sustainable landscapes, and their approach to collaborative communication and problem solving. Ask them to be very specific. How do they communicate when a surprise happens with the design, or how their budgets work and whether or not they are complete and flexible numbers? No bait and switch. How do they create and maintain trust? They should describe how they install soil food web systems and how they know if it will be successful. Specifically, how do they write plans and provide information that can help the community with maintenance. How well do they understand change management and stakeholder engagement so that you know they will help you build trust with the community and decision makers that are taking a risk in doing something new by building a food city? Of course, ask for references and work examples to prove what they say. They might not have direct experience with your plans or designs, but look for innovative projects that they have done that required them to try something new, to be creative and collaborative to take a shared risk with a client. This will tell you that they are good problem solvers, are creative, and are willing to put in the work to figure out something new that has not been done before.

A few questions to consider upfront:

- Will this be a bidding-based process or volunteer/donation-based process?
- What is the advantage of volunteers vs. specialty contractors—are their items that require special permits such as a deck or trellis that requires structural engineering calculations?
- Evaluate contractor-based construction versus volunteer-based construction process.

Contractor Based: Construction Documentation and Bidding Process

The steps to consider in developing the drawings of the edible landscape for a contractor based process include documentation for the permit and bid packages. For projects that are incorporating roof gardens, streetscapes, and technical elements that require more discipline integration, this process is the best one to use and often required for city permit process. In constructing the edible landscape, it is important in working with stakeholders and contractors to setup communication protocols upfront in the design process. This will ensure that the delivery at project turnover to the client for the ongoing operations has met the vision and goals set forth in the beginning of the process.

Volunteer Based: Design Build Construction Process

This process can also be design build to keep costs down and usually is for volunteer-oriented farm construction projects. These projects typically have included more do-it-yourself types of elements to accommodate this type of process. Much of the design or construction can be in-kind type of donations in some situations. The steps to consider in developing the drawings or sketches of the edible landscape for the design build volunteer process consist of providing enough detail that it is easily able to be figured out in the field. A leader with both design and construction knowledge is necessary, as well the ability to understand the functioning of the systems and work with the stakeholders to maintain the vision throughout the construction. Again, setting up a communications protocol up front in the design process and allowing for transitions of leadership will provide for a more successful project.

Understanding the projected timeline and schedule and how the construction implementation will be affected by decisions are important qualities for the team no matter which construction process is followed. Typically the Design Build process is done with a combination of contractor and volunteers or it is led by a volunteer who comes from the building industry and is aware of the construction issues and process. We have worked on some projects where our contractor network resulted in material or equipment donations.

The Permit Process

The permit process varies city to city, state to state. For example, animal husbandry is not allowed in many cities. Check your local town or city zoning laws on goats by contacting your city planning and building departments. Many cities have ordinances or codes in place that limit the type or number of animals you can raise on property within city limits, especially with respect to farm animals such as goats. Others allow you to keep goats but might require you to get a livestock permit through the city office.

The first thing to understand is what type of permit is required based on the functions the food landscape will incorporate into the site plan design. A landscape on land will have different requirements than one on a roof top. A landscape in the street's public right of way will have different requirements from both of those. Start with one department and then work through others. Some cities now have an overall department such as a Department for the Environment that have been put in place to handle overlapping and overarching issues such as urban agriculture. When in doubt, start with the planning department and then proceed onto other departments that are affected by the design proposal. The mayor's office is also a place that may yield results for what you are looking for.

Incredible Edible House Idea Prototype

Commissioned in 2009 by *The Wall Street Journal* to design *the most energy efficient house they could imagine,* Rios Clementi Hale Studios developed the Incredible Edible House (Figure 4.19). Its vision took into account onsite food production to reduce the food shed footprint for a family of four, and included the food shed as an elemental system of the live/work house integrated with other systems for holistic, energy efficient, sustainable living. The systems include a power system created by harnessing solar with a photovoltaic awning

Figure 4.19 The Incredible Edible House and its components to meet a family of four's food and sustainable living needs.

and harnessing wind energy with vertical axis turbines, the incorporation of a rooftop rainwater harvesting system for water conservation and use (Figure 4.20) and technology systems that monitor and think for systems efficiencies throughout the house. The house can generate enough energy to be self-sufficient and remain off the grid through these networked systems. Other considerations were taken for efficient passive design systems like cross-ventilation for temperature control and movable partitions for manipulating spaces

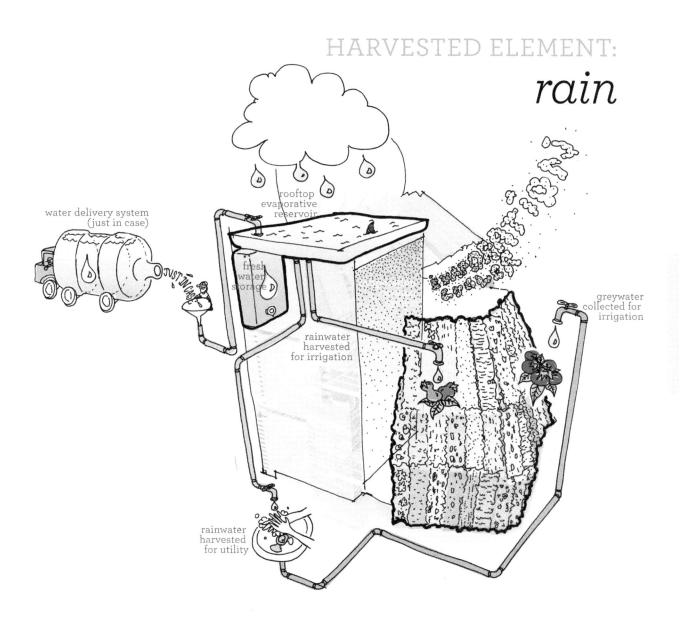

Figure 4.20 The harvested rain system of the Incredible Edible House.

to provide for the changing needs of the occupants. The prefabricated nature of the design also lends to energy efficiency by limiting the number of construction defects that can lead to wasted energy use and its modular nature provides for more efficient use of materials and building footprint.

The prefabricated structure has a hydroponic skin capable of growing chickpeas, tomatoes, arugula, green tea, and whatever produce fits the climate and season. The façade both feeds the house and provides insulation, and is considered to be a more efficient system than many traditional building materials (Figure 4.21). The hydroponic shingles use both harvested water and graywater recycled on site for crop irrigation. Water supplies can be replenished with a water delivery system if there is insufficient rain (Figure 4.22).

Another feature of the house is the height to footprint ratio—it is taller than most traditional houses, lending to a denser development and more efficient energy use. The vision includes the idea of community living by grouping several houses together in a variety of ways to meet orientation, topographic, or high-density considerations. All of the technologies currently exist to build this prototype and the firm is seeking a building partner to create a full scale prototype. They are currently conducting experiments in their office to determine the least amount of space that is needed to cultivate produce.

Figure 4.21 The Idea House is a prefabricated to reduce the eco-footprint.

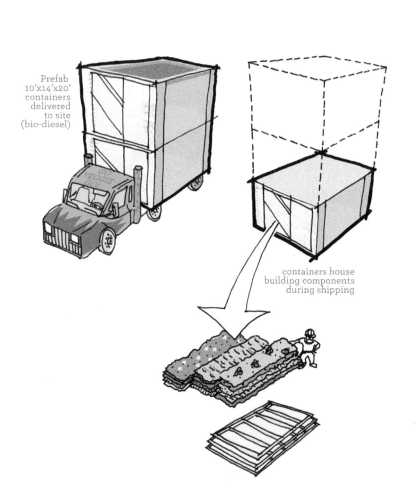

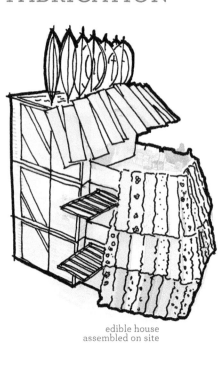

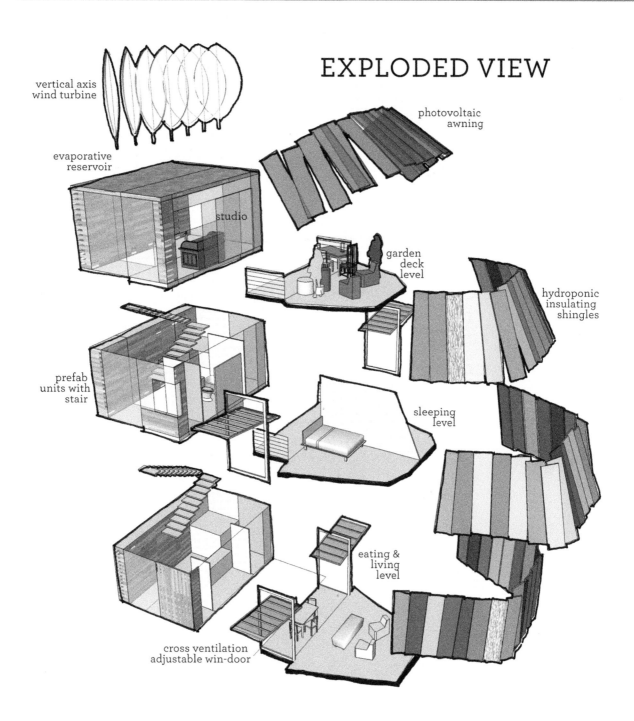

Figure 4.22 An exploded view of the parts that create the Edible House.

Design Team:

Rios Clementi Hale Studios, principals Mark Rios, FAIA, FASLA, Julie Smith-Clementi, IDSA, Frank Clementi, AIA, AIGA, and Bob Hale, FAIA, Los Angeles, California

PART 2
CONSTRUCTION INTEGRATION

An Overview of Urban Ag Construction Methods, Techniques, and Terminologies

Although many designers and planners might not become urban farmers themselves, it is important to have a working knowledge of the construction methods, technologies, and terminologies if you are designing and planning these landscapes. In particular, many of these landscapes are becoming multidisciplinary projects and, thus, it is necessary to build a "design bridge" between urban agriculture and those who are involved in the development process. The intersection between food, design, and community requires a working knowledge of methods, technologies, and terminologies for each in order to build a successful productive landscape.

There are a variety of construction methods and techniques available to select from in the refinement stage of the design process. There are also a variety of growing methods to consider as it relates to designing the landscape framework and green infrastructure elements of the project into the permit drawings.

The following terminologies represent the major methods and techniques being utilized today. Certain methods and techniques will relate better to some typologies and site-specific context over others. It is important to understand and evaluate the opportunities and constraints for building and operating the landscapes in order to finalize the elements or system connections that fit best for the project. Many books delve more deeply into the technical aspects of these methods and techniques from a gardening and farming point of view such as *The Essential Urban Farmer* or *Toolbox for Sustainable City Living*, but the following overview provides a framework of technical data for designers and planners to better choreograph the process. These brief definitions can be used to help determine the selection choices available in order to design the agricultural system that works best with the project typology but also will best fit the future maintenance and operations plans for the project.

Key Urban Farming Basic Definitions

- **Traditional agriculture**—Growing food traditionally applies to any farm that is based on the current rural farm model. Traditional agriculture farming uses a lot of land area and is based on row crops laid out in a single straight line with paths on each side. These paths can become compacted by the foot traffic that occurs along the row. Seeds are sprinkled from a packet in a singular row or planted as seedlings from a greenhouse. The entire area is watered and fertilized, not just the planted row. Crops are exposed to winds and sun. This type of planting method is generally wasteful of space, and requires a tremendous amount of mulch in order to conserve water. Typically, a 100-foot row can take up approximately 300 square feet of space. The expected yield of a crop such as carrots in this scenario would be about 100 pounds (Markham 2010). In urban areas where land is at a premium, this type of one-dimensional farming is not the most feasi-

to the land philosophy. As little as one-tenth of an acre has been determined to be the amount of land required for a small family to be self-sufficient. By expanding to a quarter of an acre more biodiversity can be achieved including more small animals such as chickens and pigs. By expanding to half-an-acre large animals can be added such as two to three goats or one cow.

Many of these landscapes are an endeavor that requires the entire household to contribute to the production and harvesting success. A project that exemplifies this movement of food security in your own backyard is a tenth of an acre, micro farm, and sustainable resource center in Pasadena, California, by the Dervaes family. They document their journey on their website and daily blog, which is a great resource for others who wish to create their own edible estate or urban homesteading environment (Path 2012).

ESTIMATING HARVEST

For a quarter-acre of land, while it is difficult to say exactly how much food you can grow since it depends on choices selected, the climate, how long the growing season is, and how intensively you choose to grow the crops, here is a ballpark number that is possible to achieve:

- 50 pounds of wheat
- 280 pounds of pork
- 120 cartons of eggs
- 100 pounds of honey
- 25 to 75 pounds of nuts
- 600 pounds of fruits
- 2,000 + pounds of vegetables

(Source: Madigan 2009)

- **Edible estates and backyard farming**—Growing food in your front yard or backyard to supplement your diet with fresh seasonal food. Sometimes a number of backyards are aggregated into a community-supported agriculture enterprise that takes care of the garden and markets the food that is grown as a business. Mostly, this represents homeowners.
- **Container farming**—Growing food in containers. In tight urban spaces with no soil farming can be achieved by using containers. Containers can be large, modular, stacked vertically, or small grow bags. You can grow as much as 16 pounds of tomatoes from one container in one season.
- **Animal husbandry**—The breeding and raising of livestock, animals such as chickens, goats, sheep, pigs, and cattle (Figure 4.26), for food or products. In urban environments, this practice may be somewhat limited in scope and is more often on the peri-urban outskirts of towns. In many cities and towns there are actually laws against raising chickens or bees within city limits.

Figure 4.26 Chickens and bees are the most prevalent animals farmed in cities, but the practice is expanding as local regulations are being changed to promote urban agriculture.

- **Community supported agriculture (CSAs)**—Over the past decade, CSAs have become a popular way for consumers to buy local, seasonal food directly from a farmer. How it works: The farmer offers a certain number of shares to the public. The share typically consists of a box of fruit and vegetables in season, but other farm products may be included. Interested consumers purchase a share, also termed a *membership* or a *subscription* and in return the consumer receives a box, bag, or basket of seasonal produce each week throughout the farming season.

- **Forage farming and urban foraging**—A recent food trend is finding wild foods and edibles in your local woods, neighborhoods, backyard, and city sidewalks. Whether this is an aspect of our sophisticated palate searching for the fresh and new or a useful way to put free food on your table in hard economic times, foraging of wild foods is now going mainstream. What used to be the way of our ancestors in learning about what is edible and what is nonedible, foraging went out of fashion after World War II. Finding wild blackberries, purslane, or miners lettuce on a local trail or pulling chickweed or dandelions from pavements or vacant lots is now an urban pastime—another sign of produce for the people. The trick with foraging is to not only know what you are looking for that is safe to eat but to leave enough of the plants to regenerate when picked. (In my neighborhood, we have blackberries along the local creek that children love to harvest in the summertime, as well as a number or fig, apple, and cherry trees in the local parks that people can harvest at will.)

- **Biointensive farming**—This is an organic farming practice that fosters healthy soils, conserves space, requires low inputs, focuses on maximizing yields from minimum surface area, and increases sustainability and overall health of the ecosystem. It is a closed system with the goal of promoting long-term sustainability. Many of the techniques were present in the agricultural practices of the ancient Greeks, Chinese, and Mayans. It is particularly effective for small farms and gardens.

Science Barge, Yonkers, New York

The Science Barge is a demonstration greenhouse situated on a barge (Figure 4.27) in the Hudson River that seeks to exhibit sustainable methods of energy generation and urban agricultural practices. At 130' × 40', the barge accommodates two greenhouses employing recirculating hydroponic and aquaponic systems, five wind turbines, rainwater harvesting system, and two solar arrays. When necessary, carbon-neutral biofuel is used. Enough power is generated to provide for the barge's operations. Originally developed as a prototype for rooftop farms, the river conditions aptly mimic what could be expected. Being located on the Hudson River, the Science Barge is ideally located to collect unobstructed sunlight and benefit from the winds that come down the river.

The primary focus is to teach sustainability through urban agriculture and is a model that can easily translate to other urban situations like rooftops. The driving principle presented is that by bringing food production in to the city, energy consumption associated with transporting food from rural to urban areas can be minimized, farmers could more sustainably manage their farmlands, and the produce consumed would be of a higher quality and nutritional value.

The food production on the barge operates with no carbon emissions, no net water consumption, no waste stream, and without pesticides. The crops produced include tomatoes, melons, greens, lettuces, and peppers and is grown with seven times less land and four times less water than field crops. Produce from the Science Barge is harvested every two weeks and is sold at the farmers market in Yonkers; any unsold produce is donated to a food pantry.

The Science Barge has developed a working relationship with local school groups (Figure 4.28) and summer camps and hosts daily educational programs from April to November. The curriculum is adjusted to the appropriate complexity for the guest age range, which can run from first grade to college level. Topics addressed include the Hudson River Estuary, renewable energy, pollination, plant life cycles, and oyster gardening. There is also a workshop series, Art and Science Sundays, which aims to relate the interactions of the two subjects in a fun and accessible manner.

Figure 4.27 The Science Barge is a floating learning center in New York City.

The barge is a working laboratory, where visitors can walk through, see the systems processes as they're functioning and be given clear, concise explanations. Steps within the cycle that are not included on the barge, like bees and other pollinators due to sting potential, are carried out by the Science Barge crew and portrayed through visuals and presentations. The barge is open to the public on weekends.

Figure 4.28 Educational programs accommodate a diversity of age groups to teach stewardship in a unique way on the Hudson Estuary. Hydroponic systems are among the technologies employed by the barge to grow food as efficiently as possible.

Design Team:

NY Sun Works Center for Sustainable Engineering

Operations Team:

Groundwork Hudson Valley Sun

Urban Agriculture Resource Issues That Affect Design Choices

Many farming issues affect design choices. They should be understood before starting up the design phase of a project. Understanding key farming issues is highly beneficial before planning and selecting the project site as well.

- **Solar orientation**—More than any other characteristic, the amount and exposure of sunlight is a critical factor in designing the layout of an urban agricultural landscape on land, on rooftops, or on vertical surfaces. Doing a site analysis that factors in sun and shade diagrams is particularly important in urban areas where buildings can block the sunlight. Food plants that require full sun for best performance need at least six hours of full sun each day in order to produce. Food crops that require partial sun need to be kept out of full sun so that they are not exposed to harmful rays that would burn them or to understand that they need to be behind or under crops that might protect them. It is important to understand what type of crops are expected to be grown when selecting a site or laying out a food landscape early on, so discussions with the stakeholders and local farmers and gardeners is an important resource to consider when evaluating the solar exposure.

- **Circulation and access**—When laying out an urban ag landscape it is important to be able to allow for universal access and enough area along pathways to care for the plants and to harvest them. Paths between planting beds should be at least 3 to 5 feet wide. Accessibility should be universal, though in some areas of severe topographic regions an area can be set aside for universal access.

- **Material selections**—Use of local materials that are readily available as well as the salvaging, recycling, and repurposing of materials are important material considerations when designing food landscapes, especially for projects with lower budgets relying on volunteers and donors. In organic farming, it is important to select materials without any chemical preservatives. Sustainably harvested wood, paints, and sealers that have no volatile organic compounds (VOCs) should be utilized.

- **Water availability**—All vegetation requires water; the question is how much is needed and what is the access to clean, fresh water. Most vegetable beds require at least 10 minutes of water per day but this will depend on the crop and other aspects of the layout. Generally, using rainwater as an irrigation source for vegetable gardens through drip irrigation or subsurface irrigation techniques is considered safe. All vegetables should be cleaned with potable water once they are harvested. Spraying non-potable water on leafy harvestable vegetables such as lettuce or broccoli has risks, so is not advisable unless it is being used for fruit trees or beneficial plants. A question to consider up front is: will the produce be harvested from rainwater-based irrigation and stored on site or will irrigation be tied to the domestic water system? Some irrigation systems have a dual system for water delivery with rain water harvested water used for part of the year and potable water being used for the rest of the year such as California, which has 6 months of minimal rain in the summertime. The availability, the type, and the amount of water are critical factors in designing the layout and location of an urban agricultural landscape.

- **Soil health and nutrients**—Along with sunlight and water, cultivating soil health is extremely important for urban agriculture. The soil food web is the mix of billions of microorganisms and little creatures that live in the soil as a complete ecosystem. It is a complex set of food chains and symbiotic relationships that include plant roots, water, air, soil composition, and soil texture. All of these elements are important to be present and balanced to ensure the plants in any ecology have available to them the nutrients and water needed. Managing the soil ecology is essential for successful sustainable land management due to soil helping manage nutrients, pests, weeds, water, and the cycling of waste.

- **Bioremediation**—This is a process that can be used for removing pollutants and contaminants from the soil though the use of microorganism metabolism. Phytoremediation and composting are some of the methods to consider if there are metal contaminants present in the soil. Soil tests should be done at an early stage to understand if this will be an issue for the urban agriculture. Some heavy metals such as cadmium and lead are not easily absorbed. In this case, using plants such as sunflowers to absorb the toxins and then removing them from the site can help to mitigate toxins. There are certain types of fungi that can mitigate other toxins through a process called mycoremediation but the right one needs to be selected for the specific pollutant. Soil experts can assist in the recommendations and monitoring if the soil falls into this category. Many urban ag farms use raised beds to protect themselves from the contaminants with a protective barrier since bioremediation can be a time-consuming and costly process.

- **Composting and waste management methods**—The most sustainable option for agricultural composting is to create a closed-loop system for organic green and brown waste into an on-site composting system that, once broken down, becomes a rich brown soil building compost to put back into the garden. Depending on the type of operation, you also may be incorporating animal manures (such as from chickens) into the composting process. Many states have their own rules about agricultural composting, so you need to check the codes and ordinances before you determine what type of operation is used. A compost area may include a bin or a series of bins that help the biodegradation process, which must be kept to certain temperatures in order to achieve the right results. Smaller gardens such as school gardens may consider using worm bins, also known as vermicomposting, for composting, both from a space saving and an educational advantage.

One waste management issue for many edible landscapes is having a surplus of food. There are many options for donating a surplus that goes back into the community to the needy, through food banks, charitable organizations, school lunch programs, and more. So do the homework to understand the potential partnerships that might be available as a resource of managing excess food production.

COMPOSTING FACTS

- More than 67 percent of the municipal solid waste produced in the United States, including paper products, is compostable material.
- A typical US household throws away an estimated 474 pounds of food waste each year. That equals about 1.5 pounds per person a day.
- Food scraps generated by all households in the United States could be piled on a football field more than five miles (26,400 feet) high. (Source: urbanbackyardedibles.com)
- Composting cuts down on waste sent to landfill, which is an environmental advantage and it creates healthy soil that becomes "food" for your garden.
- Up to two-thirds of most household trash can be composted.
- It is estimated that about one-quarter of the United States methane emissions are due to organic waste rotting in landfills.
- As long as factors such as proper aeration and proper combination of ingredients are met, compost piles can thrive in temperatures above 50 degrees Fahrenheit and do not need to be hot, though hotter piles do decompose faster.
- Compost cleans contaminated soil.
- Compost reduces the need for synthetic fertilizers and other forms of store bought soil so it saves you money.
- Chicken manure is a great fertilizer for gardens but also can be used to activate compost to speed up the process.
- Composting is incredibly low maintenance.

(Source: EPA.gov)

- **Seasonal climate impacts**—Extending the season. In its simplest form, extending the growing season allows one to grow warm season crops longer if your climate does not allow for the adequate amount of growing days. Season extension could range from cold frames to heated greenhouses in areas where snow and cold would not allow for growing in the ground. These solutions would require additional space needs, so it would need to be planned for during the design stages of the project.

- **Proposed maintenance and operations logistics**—Items that can affect the design considerations for maintenance and operations logistics not counting the composting and waste management stream, include tool storage, quantity of tools, incorporation of a greenhouse for seed starts and seasonal extension of productive vegetation, work benches, drying tables, and seed storing if seeds will be collected, office for operations and volunteers, distribution impacts, will produce be carried off, taken to market, sold on site, shared/traded—understanding how food will be distributed may have spatial impacts on the design layout.

- **Organic, semiorganic, and nonorganic landscapes**—Organic refers to the way farmers grow their food, and are designed to encourage soil and water conservation and reduce pollution. Organic landscapes do not use chemicals to control weeds or prevent disease. Most people agree that organic produce has much better taste than industrialized produce. There are currently scientific debates on which provides the better nutritional value but the metrics seem to be inconclusive on that issue. Organic and natural do not mean the same thing. Only foods that are grown and processed according to USDA organic standards can be labeled organic. Non-organic landscapes are not a sustainable manner to farm. A farm can follow organic practices without being officially certified.

Water Management Techniques That Affect Design Choices

- **Aquaponics**—As noted previously, this is a recirculating environment with fish and plants in a nature-mimicking process. It requires the specific knowledge on how to keep the constructed or cultivated ecosystem balanced with fish, plants, microbes, and worms. The waste product of the fish provides nutrients for the plants which in turn filter the water that the fish are living in. It is important to talk with the aquaponic expert and understand the ongoing operations that affect budget and staffing. Ecoliteracy as an objective could also become a part of this management technique since it requires training of the people who will operate the system.

- **Hydroponics**—In this recirculating water system, because the nutrient water supplies the food to the plants in place of soil, it requires the specific knowledge on how to create a mix of nutrients to deliver the right combination of nutrients required by the specific plants. One issue is that the nutrient solution eventually becomes toxic to the plants and will require disposing of, potentially a toxic waste issue. Should this occur a new nutrient mix would need to be created. It is important to talk with the hydroponic expert and understand the ongoing operations that affect budget and staffing. This technique could also be part of a green job/educational training program.

- **Rainwater harvesting**—In urban areas, water rates are typically higher than the reduced rates of rural farms, and since agriculture requires a lot of water, catchment of rainwater is an effective way to promote water conservation and reduce the amount of water required for irrigation. There are a variety of ways to harvest rainwater from small-scale rain barrels to catchment systems such as cisterns. Understand which ones are practical for the project and what its spatial and physical requirements are in the early stages of design.

- **Reclaimed water**—Harvesting water from household wastewater, also known as graywater, can be restricted in many cities. However, there are technologies such as constructed wetlands and other means to clean collected kitchen, and laundry wastewater. These treatments need to be considered when developing the spatial needs of the project, as they do require land areas and swales for treatment zones and some more sophisticated systems require pumps and filters. Never apply raw graywater directly to plants you plan on eating raw or plants whose leaves or fruits are within reach of it. Graywater is good for watering orchards, shrubs, and compost piles. There are a number of books on the various methods to consider if reclaimed water is going to be integrated into the design.

- **Irrigation technologies**—Drip irrigation is more efficient than a hand-watering system. Drip irrigation can be through emitter lines, a soaker hose, spot emitters, micro spray emitters, and T-tape. Drip systems can conserve about 50 percent of water used by other methods. Installing an on/off valve at each planter bed is an effective way to control for the variable watering needs of crops selected.
- **Living machines/black water**—The technologies for living systems are extremely expensive and will likely not be used for most urban agriculture landscapes in the near future. Should the project be extensive enough to be considering the use of living machines for the project, it would be a way to design systems that are more regenerative and balanced for all of the open space and building systems. Living machines such as the one installed at Oberlin College provide a great educational resource to promote innovation technologies into sustainable landscapes.

Considerations for Animal Farming in Urban Environments

- **Beekeeping**—Bees are important for improved pollination and often considered the gateway urban farm animal. Beekeeping is fairly easy to do and bees are easy to care for. Bees also provide honey, wax, and pollen that can be either harvested for personal food use or processed for market sale production. The do have spatial requirements to consider for beehive location and honey harvesting purposes but are even suited for rooftop locations. Water source availability, storage space, protection from ants, or passersby are also considerations for beehive layout.
- **Chickens/ducks, turkeys/ rabbits/goats**—In urban environments, animal husbandry of any sort is subject to animal ordinances so the first thing to do is check on the local regulations before planning for them in the design. These smaller animals provide multiple uses for a food landscape. Chickens provide eggs and fertilizer while and goats provide milk for cheeses as well as weed management. Both can also be harvested as food themselves if allowed. Key items to consider in design are spatial requirements for habitat (Figure 4.29) and relationships to food landscape use, water accessibility, health requirements related to breeding and/or slaughtering, and metal contamination in urban soils that may limit their foraging abilities (Figure 4.30).
- **Fish**—These can be raised for sustainable farming or incorporated into an aquaponic system. However, most aquaculture farming is not especially suited in cities except in open space park types of land uses. These types of uses are not typical, but they are slowly starting to appear in some cities. The more common use for fish in urban environments is as fertilizer, in the form of fish-derived soil additives. A project that recently won a Buckminster Fuller Challenge Grant called "The Plant in Chicago" is creating an interior aquaponic system in a converted warehouse as a vertical farm that will benefit the local community on a year round basis (www.plantchicago.com).
- **Other larger animals**—Because sheep and cows require a much larger amount of land per animal to raise them these animals will generally not be incorporated into most urban farms or urban ag landscapes within a city.

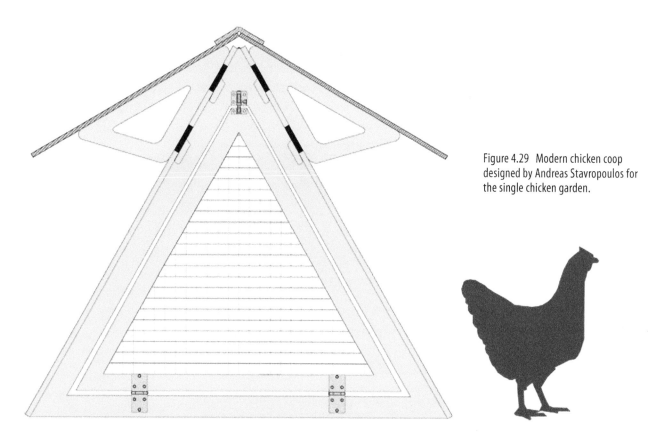

Figure 4.29 Modern chicken coop designed by Andreas Stavropoulos for the single chicken garden.

Figure 4.30 Chicken coops come in a variety of sizes and present a chance for visual aesthetics that meet function. This coop was designed for five to six hens.

Emerging Products and Technologies

Technologies are constantly being improved or invented when it comes to urban agriculture, so research is an important part of the design and planning phase to see what types of high-tech and low-tech products or inventions might be harnessed for specific projects especially as they relate to physical layout and space requirements. Here is a short list of some basic products to consider:

- **Vertical planters or walls**—There are a variety of products for making better use of vertical surfaces for food landscapes in urban environments especially in high density areas when horizontal surfaces are not available for production type of use. Movable solutions such as hanging planters made of recycled felt or fabrics (Figure 4.31); or more permanent solutions such as modular living wall system products to consider when expanding the available real estate for a food landscape by going vertical (Figure 4.32). Issues to consider with selection of product are the growing mediums, which can either be soil based with drip or hand-watered irrigation or hydroponic based; the size constraints, water accessibility, and orientation to sunlight.

Figure 4.31 The Woolly Pocket garden at Miller Creek Middle School provides a way to expand the square foot garden surface vertically as well as buffer or baffle a view or building in a more friendly and functional way.

Figure 4.32 Vertical surfaces can be from salvaged wood and use a combination of sedums and herbs for a simple way to grow some garnishes.

- **Modular roof planters**—Several recyclable products available for rooftop gardens can be used for urban ag landscapes that can be purchased and networked together.
- **On-site recycled materials and urbanite**—There are a number of ways to use recycled or salvaged materials, including bricks and stone or repurposed wood for raised planters or fences. *Urbanite* is a word used to describe broken-up concrete that can be stacked in blocks to create a raised planter bed or low wall.
- **Intensive and extensive self-watering planters**—A number of modular products similar to green roof products can be purchased for both rooftops that are intensive or extensive with self-watering and self-daring capabilities built into the product to increase water retention and draining.
- **Urban ag rooftop soil mixes**—With the increase in urban farms and community gardens on rooftops, which requires soils to be more lightweight to reduce structural loads on a building, a number of soil manufacturers have come up with soil media suited for urban agriculture needs for rooftops and vertical walls.
- **Landscapes as building skins**—This is similar to the vertical category, and there are several product options to select from.
- **Solar technologies**—A few new technologies are combining solar products with urban agriculture planter products on rooftops as a dual-system product that synergizes with the other for maximum benefit.

Banyan Street Manor, Honolulu, Hawaii

Located in the Kalihi neighborhood of Honolulu, Banyan Street Manor is a low-income, medium-density family housing community. It was originally built in 1976 and previously managed by the Hawaii Public Housing Authority, but in 2011 the property was purchased by the Vitus Group, a developer that specializes in low-income housing, with a focus on green design and construction, as well as community revitalization.

After purchasing the development, the company invested more than $3.1 million in improvements, including a solar power system, solar water heating system, a cooling white roof, and a group of living walls that are the largest in the state. One of the most community beneficial improvements is a rooftop farm, which is the first USDA-certified organic rooftop farm on an affordable housing project.

Completed in April 2012, the semi-intensive rooftop farm (Figure 4.33) was designed by 1st Look Exteriors, who also manage the farm in cooperation with Vitus Group. Using 2,000 square feet of growing space and planter equipment from Green Living Technologies International, the farm produces more than 20 types of crop plants, including tomatoes, eggplant, lettuce, strawberries, green beans, and a variety of Asian herbs.

Nearly half of the harvest is going to the residents of Banyan Street free of charge, while the balance portion of the produce will be sold to local grocery stores to offset labor costs to cover the farm's operational overhead. The farm helps accomplish Vitus Group's mission of increasing sustainability, promoting healthy living, and enhancing the sense of pride and community for their residents.

Figure 4.33 The Banyan Manor affordable roof farm with a view toward downtown Honolulu.

Design Team:
Client: Vitus Group
Designer: 1st Look Exteriors

Intensive Planting Methods to Consider

- **Intensive farming**— As previously mentioned, growing food intensively is an efficient and eco-effective way of growing food. This includes the laying out of crops so that no soil is exposed to sunlight. The plants' leaves touch and provide a canopy of living mulch over the soil so that moisture is maintained at the plant roots. This results in more efficient water conservation and increased soil health. This type of planting results in a higher yield of produce within a much smaller area because of the density of the biomass and is extremely useful in urban ag landscapes as a means to maximize the area of land available. There are a variety of intensive methods to select from which follow.

- **Calorie farming**—Concentrating on growing the highest calorie-laden crops for a complete diet to live on in the smallest area possible is called calorie farming. This includes focusing on special root crops that are nutrient rich and calorie dense, such as potatoes, sweet potatoes, leeks, garlic, parsnips, Jerusalem artichokes, and salsify. Combining root crops with leafy greens high in vitamins and minerals along with fruit will provide a diet that meets a person's caloric needs.

- **Companion planting/ beneficial vegetation compositions**—Companion planting focuses on combining plants that create a thriving mini ecosystem with beneficial relationships. This technique draws a diverse insect population to the garden by using plants of many types and colors that flower all season long. These plants also provide a place for insects to drink water and be protected at night. These actions will support a balance of beneficial insects that prey on insect pests and pollinate the crops. Also, choosing strong-scented plants like marigolds and chives will help repel unwanted insects.

- **Square foot method**—Intensive agriculture method that recommends using a grid of squares dividing every square foot of planting surface area into a number of subsquares appropriate to the spacing of the crop that is being grown. This method works best on a small size garden of 200 to 400 square feet. It uses a combination of organic gardening methods such as compost, densely planted beds and biointensive attention. The phrase "square foot gardening" was popularized by Mel Bartholomew in a 1981 Rodale Press book and subsequent PBS television series.

- **French intensive**—A planting method that was started in the 1890s in Paris using less land and water to maximize production. It uses organic compost for the garden and is densely planted. Each bed is mounded typically 5 to 6 feet wide and 12 feet plus long, with 3-foot paths between them. It utilizes a double digging method by layering organic fertilizer to prepare the garden which is more time-intensive but provides better drainage and more surface area to the plant bed. There are claims that it produces up to four times as much produce and half as much water as traditional methods.

- **Biodynamic**—An organic form of farming that emphasizes the holistic relationships between the soil, plants, and animals as a self-sustaining system. It emphasizes a sustainable approach to agriculture that focuses on maximum yield in a minimum area of land. It integrates the cultivation of the land with farm animals through manures and composts, as well as fermented herbal compost additives. An astronomical chart guides the timing for planting and sowing. This method was further developed by John Jeavons and Ecology Action into an 8-step food raising process. Love Apple Farms in Santa Cruz, CA, is an example of this type of method.

APDW staff created these colorful signs for the edible garden at the VF Outdoor campus.

Leafy greens ready for harvest at VF Outdoor.

Resources

Bartholomew, Mel. *Square Foot Gardening*. Emmaus: Rodale Press, Inc., 1981.
Carpenter, Novella, and Willow Rosenthal. *The Essential Urban Farmer*. New York: Penguin Group, 2011.
Kellogg, Scott, and Stacy Pettigrew. *Toolbox for Sustainable City Living*. Brooklyn: South End Press, 2008.
Madigan, Carleen, ed. *The Backyard Homestead*. North Adams: Storey Publishing, 2009.
Markham, Brett L. *Mini Farming: Self-Sufficiency on 1/4 Acre*. New York: Skyhorse Publishing, 2010.
Path to Freedom. "Urban Homestead." 2012. http://urbanhomestead.org/.

A corten planting bed at the VF Outdoor edible garden.

CHAPTER 5
Lifecycle Operations

Die Plantage at BUGA 2005, Munich, Germany

Die Plantage, "The Plantation," is a three-acre public park and fruit orchard (Figure 5.1) originally constructed for the 2005 German Federal Garden Exposition (BUGA), a year-long horticultural and landscape exposition that took place on the site of the former Munich airport (Figure 5.2).

Figure 5.1 The three-acre park and orchard serves the adjacent residential community and is connected via trails and walks.

Figure 5.2 The former Munich airport was both industrial and agricultural land.

181

182 Designing Urban Agriculture

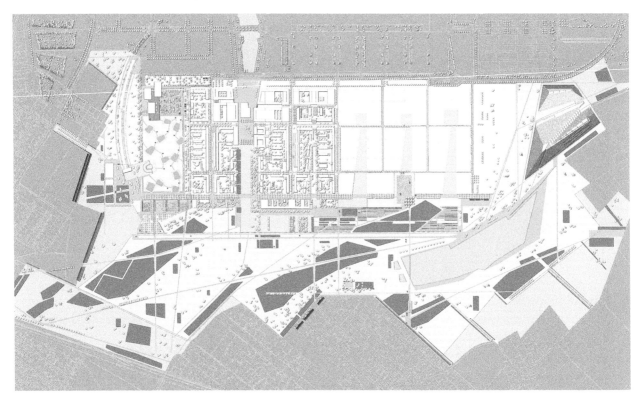

Figure 5.3 The overall BUGA 2005 master plan.

Figure 5.4 The park's original exhibition gardens were removed to create park and gathering spaces for nearby residents.

Die Plantage is part of the larger 200 hectare Munich "Landscape Park Riem" designed by Gilles Vexiard of Latitude-Nord on the former airport site as part of the 560 hectare new residential community (Figure 5.3). In 2000, a competition was held for BUGA 2005 and Die Plantage was selected for its farming the landscape idea "Sunken gardens and Plantage," conceptualized by Rainer Schmidt Landscape Architects and collaborating team members Architect Reinhard Bauer and communication Designer Axel Loritz.

It is a public space that is an example of the integration of performative landscape processes in ecological design and infrastructure. Die Plantage represents the integration of productive landscape components as a system of the local food shed integrated into the public park's open space systems. The Rainer Schmidt team was responsible for designing intensive zones as permanent gardens beside and connected to the defined park area. The designers saw the sunken gardens and orchards as a combination of ecological and sociological planning systems. Once the exposition was over, the three-acre orchard park remained accessible to the public as part of the open space of the new mixed-use development now situated on the former Munch airport site. The overall planning strategy was to provide a public park in Riem for the future inhabitants of the new residential community after the restructuring of the former airport. The garden exhibition area was structured as temporary open spaces (Figure 5.4) for the future residential areas.

The design is inspired by the *Streuobstwies*, or meadow orchard (Figure 5.5), a traditional cultural landscape of southern Germany. This idea has its origins from the walled gardens of medieval towns and villages normally surrounded by gardens.

Figure 5.5 The orchards and meadows provide a park setting year round and the harvesting of the fruit trees is by the general public and adjacent community.

Planting the fruit trees in the outer agricultural zone to surround the intensive inner urban community zone makes an ecological statement to the development (Figure 5.6).

The lowering of the orchard into a sunken zone added a climate spatial advantage for the fruit trees because of the sheltered situation and the slight difference in seasonal temperatures. The park's 137 fruit trees are arranged in a grid, originally with a surface of decomposed granite, reminiscent of a formal bosque in a sunken garden (Figure 5.7). Instead of ornamental trees, the orchard contains 16 traditional and heirloom apple, cherry, and pear cultivars used in the agricultural landscape of the region. Thus, instead of simply providing shade and form, the trees form a space that is both productive and educational. The trees were planted in decomposed granite for the year-long exposition (Figure 5.8), but afterward it was replaced with a lower maintenance crushed aggregate lawn that was more inviting for the park visitor. This technique means that the area was seeded with dry lawn, covering the ground with top soil and re-naturalizing the entire zone in this manner.

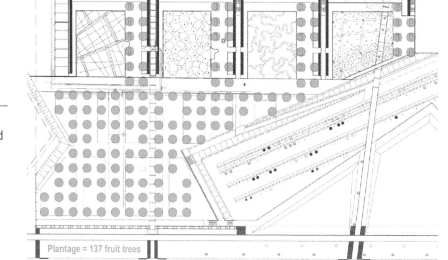

Figure 5.6 Die Plantage site plan showing grid-like arrangement of sunken gardens and orchard.

Figure 5.7 An overview of the BUGA 2005 temporary garden plan.

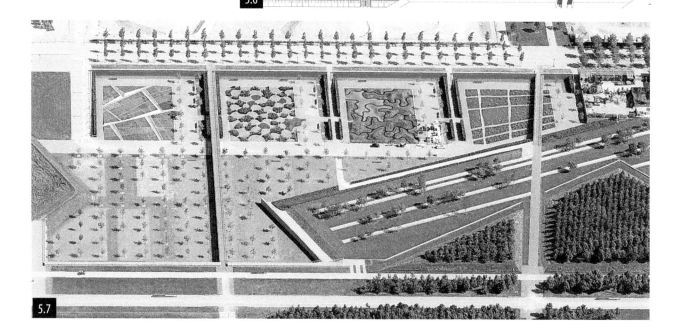

Figure 5.8 An axonometric view of the BUGA 2005 orchard and temporary garden plan.

The lawn surface leaves more programmatic cultural possibilities for future additional uses as the surrounding neighborhood evolves. The fruit trees are now harvested by park visitors and the surrounding residents of the new neighborhood. A circulation system of paths connects the park to the new community.

As far as the ecological restoration process is concerned, the team believes that Die Plantage represents and mediates sustainability values in terms of cultivating local and regional specifics: Common ground, "Allmende"-like, providing for a field of urban agriculture that combines the supply of locally grown food with the socio-spatial aesthetics of a plantage. The flowers of the trees, blooming in spring, and the fruit, ripe in autumn, ready to be harvested or picked from the ground generate an image for the new urban area that can be called a very specific "habitus" for nature and people.

Temporary and permanent gardens fit into the planning goal for combining urban agriculture into a linked open space system that serves the community in a diversity of ways. Die Plantage demonstrates that productive food landscapes can work in multifunctional public open space, provide ecological performance landscapes, enhance community activity, and provide a unique placemaking component to public space.

Design Team:

Landscape Architect:	Rainer Schmidt Landscape Architects
Architects:	Reinhard Bauer, Judith Stilgenbauer
Communication Design:	Axel Loritz

Setting the Stage for a Lifecycle Operations Approach

The term *lifecycle* refers to the useful life of a product or system. In the case of urban agriculture, it also refers to a more sustainable approach where the three main systems—ecology, culture, and economic core building blocks—are actively pursuing a balance toward a renewable systems network that is focused on providing a relationship of synergy between the community and local economy. Thus, a lifecycle operations approach is tailored as a dynamic and iterative process for managing the landscape on an ongoing and self-sustaining or regenerative manner. Building resiliency is the optimum goal.

A more traditional lifecycle approach to operations or management is a cradle-to-grave system. This is a finite system that encompasses a beginning, middle, and an end. For urban agricultural landscapes, a finite system is not the best choice, nor is it the sustainable choice. These types of landscapes require an ongoing lifecycle system that ensures project success. In order to achieve an eco-balance of renewable systems as the project progresses, lifecycle operations would be designed to allow for modification of its business structure, its technologies, its production math, and its human levels of input as required to maintain a self-sustaining project. This includes being able to accommodate a flux of the old systems with the new, and at the same time providing for systems mitigation strategies, replacing old technologies with newer, more innovative ones. It is also engaged with providing for a means to optimize the allocation of necessary inputs into all of the systems in order to continually integrate them into meeting the objectives. This lifecycle approach is modeled on continuously aligning the business and the operations management with the project's vision and the objectives. It seeks to evolve the operations as the vision evolves over time.

Having a basic understanding of what an urban agriculture lifecycle operations approach entails will begin to ensure that the lifecycle infrastructure will already be in place by day one of operations. In the planning and design phases, this process starts with an outline and a brief narrative about what the projected management and operations structure might look like. This outline should be discussed periodically with the project's team and stakeholders for feedback to keep it on track with both budget and management projections. All of the component plans that embody the lifecycle aspects of the urban agriculture landscape begin in outline format in the preliminary business plan as the project stakeholders identify and refine the project's goals and objectives. These lifecycle outlines continue to be modified as the project evolves and the design form begins to become more physically apparent. The basic lifecycle choices of the plan should be finalized with project team and stakeholders before the project is built so that a lifecycle framework is in place to ease the transition from construction to operations.

Interdependence of Maintenance and Management

The common ground between the maintenance and management of urban agriculture landscapes should be focused on pursuing, and ultimately attaining, a relationship of interdependence. A lifecycle approach is about building resilience into the productive landscape and its systems. A food landscape is a dynamic system; it is never the same from year to year. The challenge in planning for interdependence is in understanding the flexibility of the moving parts enough to allow for adaptations that will still provide for a cohesive and balanced whole. This dynamic relationship of the systems is perhaps the trickiest part of the process for urban agriculture landscapes.

Organizing the framework of the human work patterns within the landscape to provide for stacking functions or lifecycle multiplicity of use aids in establishing work patterns where people make the landscapes and gardens part of their lifestyle as much as possible, not just an extra thing to manage. When seen as a mutually responsible relationship, a strong communications network becomes the thread to harnessing the productivity, economics, and community benefits of urban agriculture. For landscapes that are nonprofit enterprises, there will be some dependence on a revolving volunteer workforce for maintaining and managing the landscape. Larger for-profit enterprises may have more stability in workforce but still need to allow for the nuances of the changing conditions ecologically, socially, and economically. Having a system in place that allows for input and output based on observation and feedback by maintenance to management and vice versa sets up a healthy role model for the enterprise. The food enterprise will tend to self-monitor and self-guide over time with an interdependent relationship in place. Management and maintenance thus become part of the fluid dynamic with each supporting and achieving mutual goals.

If your urban food system will include for-profit production on a large scale, it may be important to engage with the community on a large scale, over a long period of time, to develop the business plan to ensure it is sustainable for all involved. If a farmer creates a business plan in isolation from the rest of the local community to sell vegetables, that farmer has to be the grower, the marketing person, salesperson, administrator, lobbyist, etc. If the whole community that is interested in seeing a successful food system exist in their city comes together to create a systemwide business plan, then the farmer can grow the vegetables knowing that neighbors are working on other parts of the food system with the same goal: local vegetables.

A great example of this is the Community Table in Minneapolis, Minnesota. This group was started as a network of people interested in a values-based local food system that works together to design and determine what the business model for the whole food system could be. They are working together and building the business model based on three values that guide all of their work together: equity, transparency, and trust.

Many people work on this network to ensure it will be designed to be sustainable and thrive. Members come from all walks of life and have a variety of skill sets. It is not a network of urban farmers, but a network of interested citizens who want this system to exist in their communities and lives. The system is more likely to succeed due to the politicians, buyers, growers, land owners, processors, distributors, regulators, cooks, and businesses all adding their perspective in the design of the system to ensure that it is something they will use as it gets off the ground. This approach distributes the risk out to many because the people working together trust each other and are transparent with what they are willing to do and not do to create and exist in the system. Participants offer what they can and know specifically how they can benefit from the food system. This creates transactions and a new food economy and ecological system based on the values of equity, transparency, and trust to ensure people have control and buy-in of the new food system.

The Lifecycle Operations Sphere

The lifecycle operations sphere (Figure 5.9) for urban agriculture landscapes includes the development of a series of plans that create the network of connections among one another. These plans will become part of the overall oversight for the food enterprise. Designing a lifecycle operational plan is about developing mechanisms for creating synergies between maintenance operations with management operations in a relationship that is mutually beneficial and interdependent. This includes determining not just what needs to be done but how it is to be done and setting up an operations budget and finance plan along with the maintenance and management plans. All of these plans are, in turn, linked to the marketing outreach plan. There is a strong connection between the lifecycle operations sphere that focuses on building resiliency and the outreach sphere that focuses on longevity and stewardship through education and marketing programs.

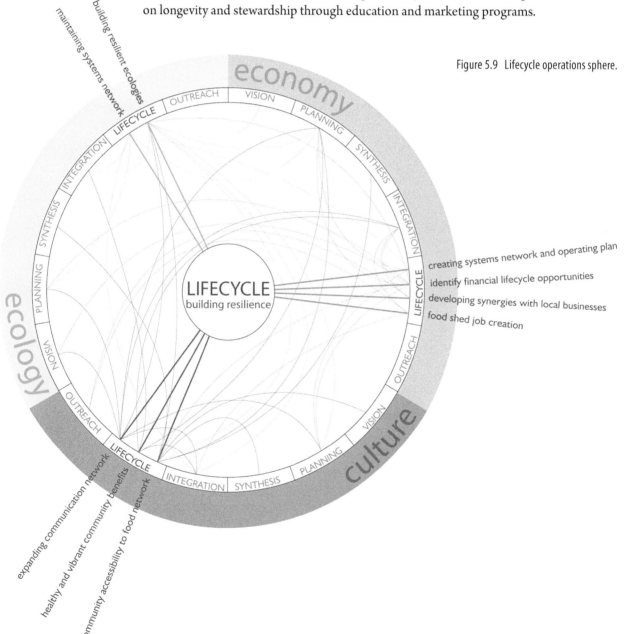

Figure 5.9 Lifecycle operations sphere.

> The lifecycle operations plan includes tasks such as identifying the projected management and operations approach, identifying the ongoing operational budget, mapping the tangibles and intangibles for the ongoing maintenance, and developing the physical infrastructure required for the management and operations of the food landscape as part of the local food shed.

Lifecycle operational considerations include items such as annual budgeting allowances for seasonal crop considerations, identifying appropriate harvesting options, incorporating green jobs such as garden coordinators, managing a seed collecting system for heirlooms, and seed storing for regenerative harvest production, organizing volunteers, providing mentorship and training programs, managing the energy flow and water systems within the green infrastructure, managing waste management through composting and harvest distribution, and many more operation considerations. This lifecycle operations process begins in the design process with the stakeholders, not after the landscape is built.

Setting Food System Maintenance Criteria and Goals with Stakeholders

How will the landscape be managed once it is built? The landscape operations approach sets a framework from the beginning of the project, so that the stakeholders have criteria for addressing the practical issues once a vision becomes reality. During the planning and vision spheres, visions and goals are established. The synthesis and integration spheres that follow begin to define the connections of the systems. The maintenance and management criteria are then developed that will best implement the food landscape's vision, mission, goals, and systems. Using this approach, the project programming includes these considerations upfront in its synthesis and integration stages so that by the time a maintenance plan is written in the lifecycle operations sphere, many of the projected maintenance and operations criteria have already been factored into the design of the urban ag landscape.

It is important to design the maintenance of the food systems (Figure 5.10) with the people who will own, rent, work in, and live near it to ensure that the maintenance plan you design is one that fits into the daily rhythms of the community. People who are responsible for the gardens can help determine what tools and tasks are needed for the maintenance, which can inform the design of pathways, community spaces for gathering buyers, cleanup ease, maintaining aesthetics, ease of moving organic materials, space for harvesting and cleaning, restrooms, and more. Every detail of how the garden will change and be used throughout an entire year will inform how modular the space needs to be to satisfy not only the needs of the workers but also the use of the community and how well it weaves into the fabric of the surrounding community. The dynamics of the garden changing can be complimentary to the way a community wants to use the space to ensure that people come to use the space and associate with their enjoyment of living there. If the community can enjoy the gardens being in their community, they will want to help it succeed for the long term by purchasing foods there, voting to spend tax dollars on parts of its upkeep, helping on volunteer days, and therefore helping to ensure its viability as a food system and increasing people's quality of life.

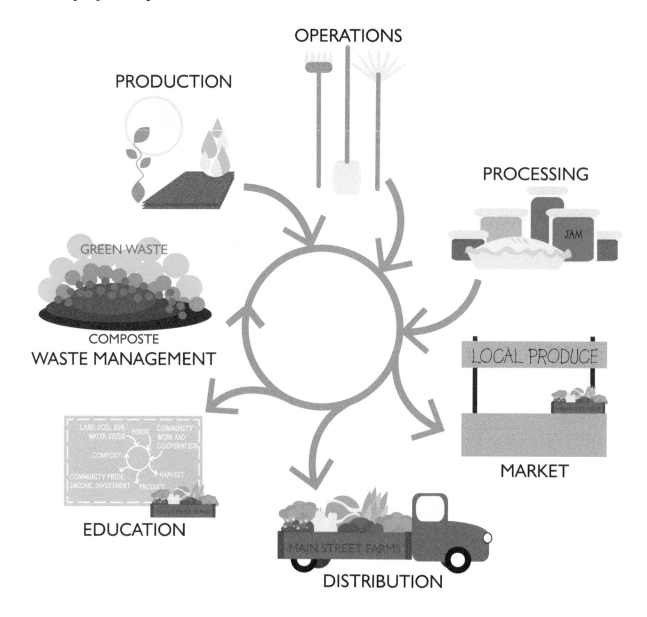

Figure 5.10 Lifecycle illustration of urban food system.

It will also be important to identify the expectations and needs of the food system stakeholders. This will help the community agree to certain thresholds that they prefer not to exceed so they remain satisfied with the project. Each member affected by the gardens will have a time when something is unacceptable about the garden, which could lead to conflict in the community. Thresholds include amount of weeds, amount of water used, space available for the community, use of pesticides, profit needed by the farmers, amount of tax dollars, and positive image of the gardens. Most needs that stakeholders have can be turned into a useful threshold by asking when will they be happy, when will they be unhappy, and identifying the line that will be crossed. In a food system that is dynamic and changing, it is important to set up expectations and processes that help the community navigate potential conflicts when a threshold of acceptance has been surpassed. During the planning stages for maintenance, you can have collaborative conversations with all the stakeholders at the same time to come to consensus on the threshold levels. In the documents created for the gardens after completion, the community can list these thresholds and agree on what measures will be taken if they are exceeded. The most important part of this document will be to have the stakeholders agree that when a threshold is exceeded, they will call a collaborative conversation to discuss the best solution for moving forward, and will use a facilitator if needed. This allows the community to focus on the solutions instead of the problem and blaming someone for it. Since everyone agreed on the thresholds, they can agree something should be done about it. The trick is making sure they are set up for success to deal with the tension of exceeding a threshold. This also allows the garden to evolve with the community after the design team has left.

Medical University of South Carolina Urban Farm
Charleston, South Carolina

The new Urban Farm at the Medical University of South Carolina does more than grow food; it serves as a living classroom that promotes healthy eating and living for both the university community and the greater public of the Charleston area (Figure 5.11). It was originally a parking lot (Figure 5.12) and slated to be a nonproductive green space, but landscape architect Bill Eubanks saw potential for something more. His initial vision of an urban farm was embraced by the university and has become a base for a variety of activities designed to promote public health and knowledge.

Figure 5.11 The MUSC Urban Farm's central area under a large existing canopy tree, taking advantage of the shade while creating a productive landscape in the place of a parking lot.

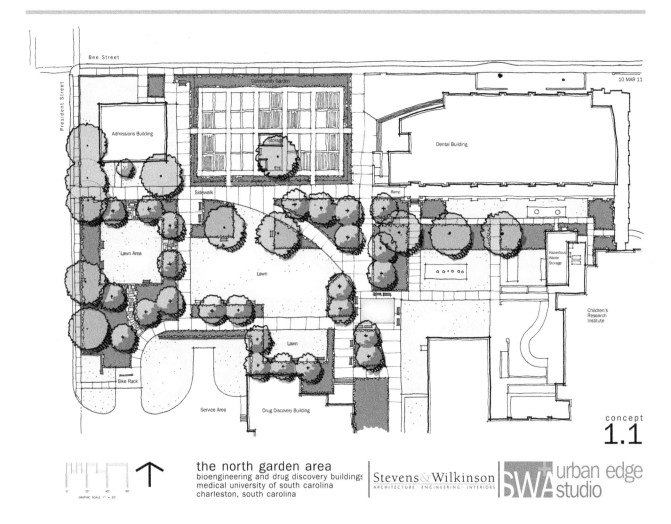

Figure 5.12 The MUSC Urban Farm was originally a parking lot between two campus buildings. This plan illustrates the scale and context of the farm.

Beyond its use as a teaching tool, the garden provides produce to the MUSC cafeteria, prepared by the Sodexo food service company, a partner in the project.

The farm uses many tools to promote healthy living. An informational handout has been created for each of the 50+ crops, explaining how to cultivate and harvest the produce, as well as its dietary nutritional information. The food landscape is accessible for strolling through, which exposes the MUSC campus to foods they might be unfamiliar with or otherwise be wary of trying (Figure 5.13). Events such as cooking demonstrations are regularly held in the living classroom to teach participants how to prepare the vegetables they are growing and harvesting.

By promoting the living classroom idea, the MUSC Urban Farm is able to bring in volunteers looking to learn and reduce the need for groundskeepers. The garden is managed by a multidisciplinary team composed of grounds crew, dieticians, and a food distribution coordinator. Work-and-learn sessions are held through which volunteers can donate their time while being educated on the cultivation, harvest, and preparation of a variety of crops (Figure 5.14). Not only can they take this knowledge home with them, but they can also take home a share of the produce. Workshops and lectures are also held regularly in the outdoor learning space.

Connections have been forged to build an outreach program with organizations such as the MUSC Healthy Charleston Challenge, Weight Management Center, and the Ronald McDonald House. Building these relationships facilitates the benefits of the garden to go beyond the campus employees and students, and into the community to promote better understanding of nutrition, eating habits, and healthier living. Harvest surplus is taken to local food banks and churches.

Lifecycle Operations 193

Figure 5.13 The MUSC Urban Farm is along a main circulation path, enabling it to reach out to students and faculty who pass by.

Figure 5.14 Built-in benches and tables add to the outdoor classroom use at the MUSC Urban Farm.

Design Team:

Architect:	Stevens & Wilkinson
Landscape Architect:	Urban Edge Studio of SW+A
Urban Agriculture Consultant:	Crop Up, LLC

Lifecycle Operations Plan Components

A lifecycle operations plan is typically composed of the following plan components:

- **Maintenance plan.** This plan is important for setting the overall maintenance framework criteria set through the earlier discussions with the stakeholders into planned tasks to be provided. It is detailed enough for preparing a projected maintenance budget and factoring the amount of labor required on an annual basis. A maintenance plan typically includes a breakdown of the estimated daily, weekly, monthly, and annual maintenance tasks. It looks at the climate factors, seasonal factors, and resource management factors. It establishes the methods for plant and soil management from attaining and monitoring the health and vitality of the produce from seeds, the crop selection and rotation methods to be utilized, and the entire growing cycles and harvesting process methods to be used. It plans the tasks to follow for the recycling and waste-stream management process in the most efficient manner, taking into consideration the local food shed opportunities within the neighborhood and city. Including feedback loops that provide a two-way system of communication is essential for successful maintenance plans.

- **Management plan.** The management plan is important for setting the overall framework for the project's oversight and personnel capacity. It consists of monitoring the food production and food system flows such as harvesting, distribution, and waste recovery systems, as well as personnel and labor considerations such as daily workforce, mentoring, and green job training programs. The management plan is typically overseen by the leadership group of a nonprofit or for-profit organization, a stakeholder advisory board, or an executive director that is hired to oversee the enterprise. This group or individual is responsible for making sure that the project's vision and goals are being met and hiring the staff needed to manage, maintain, and promote the food landscape and its programs. Including feedback loops is essential for building a sustainable management plan.

- **Finance plan.** The finance plan is important in setting up the framework for determining the financial model of the project. This will vary depending on whether the food landscape is a nonprofit or for-profit enterprise and the mission set for the landscape. It consists of setting annual budgets and identifying annual funding sources and seeks to create a plan that can be self-sustaining over a long period of time, not just for a year or two. An important financial tool is the return on investments, or ROI, on an annual or biannual basis. The ROI can help to establish metrics that provide more opportunities for ongoing funding sources. Phasing typically also plays a role in most urban agriculture projects, as do public and private grants, donations, and seed money from entrepreneurs and private investors. Evaluating the ROI is a way to measure food production in dollars per square foot and translate that number into anticipated sales and local economic terms. Setting annual budgets and revisiting the business plan on a quarterly basis is a helpful way to keep the food landscape on course and provide the necessary feedback loops in a timely manner.

- **Marketing plan**. The marketing plan sets the framework for the communication network and outreach. It communicates the project's value based solutions and relationships. It is tied to lifecycle operations in that the educational and job training programs are part of the management and operations structure. The marketing plan is part of the business plan and outlines how the vision strategies will become physical realities. It embodies the strategic planning for aligning the marketing message with the mission/vision of the urban food landscape. This would include the branding and media tools to be incorporated. Marketing, branding, ecoliteracy programs, and longevity stewardship are discussed in more detail in Chapter 6.

The Maintenance Plan Framework

The key difference in urban ag maintenance plans from standard maintenance plans in the landscape maintenance industry is due to the lifecycle approach with its focus on the networking relationships of the sustainable systems. This systems-based focus is not an industry maintenance standard. It is customary to begin the development of the maintenance plan process with the creation of a maintenance task framework outline that flushes out an understanding of how the food garden, landscape, or farm is projected to be maintained. For small food landscapes, only a maintenance checklist is needed to guide the efforts, along with a daily garden journal to record results and observations.

A maintenance framework plan would include the following components:

1. Soil management
   - Composting and compost teas
   - Mulch
   - Organic nutrient based vs. Nonorganic nutrient based
   - Soil food web monitoring and analysis

2. Produce management
   - Cover crops
   - Seed collecting and propagation
   - Crop rotation
   - Seasonal produce plans
   - Produce harvesting
   - Seasonal strategies (fabric covers, trellis, hoop houses, greenhouses)
   - Beneficial companion plants and biodiversity
   - Tools and supplies

3. Water management
   - Water conservation
   - Rain harvesting though cisterns
   - Eco-efficient irrigation
   - Graywater harvesting

4. IPM—integrated pest management
   - Setting the guidelines and protocols
   - Training
   - Identifying pests, predators, and diseases

5. Waste recovery management
   - Green waste collection
   - Green cycling into compost
   - Material recycling and reuse
   - Setting up zero waste targets and protocols
   - On-site versus off-site options
6. Harvest distribution management
   - Identifying daily, weekly, and monthly distribution options
   - Identifying partnership options with community organizations and local businesses
   - Setting up targets and protocols
7. On-site facilities versus off-site facilities for various food shed support systems
   - Tool storage and staff office
   - Energy alignment or synergies with energy corridors
   - Green infrastructure synergies
   - Seasonal extension of crops through greenhouses and hoop houses
   - Market stand for retail use
   - Delivery and transport systems such as bikes and trucks
   - Retail harvest processing such as canning or baking
8. Animal husbandry management
   - Small animals
   - Larger animals

Maintenance Mapping

One way of communicating the maintenance elements and garden calendar for a landscape is through the creation of a graphic-oriented maintenance map. This is a visual tool that can facilitate an understanding of the process and layout for stakeholders, volunteers, and staff. Mapping the tasks on a site plan can help to set up the framework for operations management to monitor tasks that need to be performed. It is also a useful tool for discussing the tasks that need to be done with volunteers, students, staff, or community members who are participating in the landscape's growth and upkeep. It is particularly useful when there is a wide diversity of volunteers or visitors who are not participating on a daily basis.

Each year of growing food can include a maintenance map to coincide with the maintenance checklist. This is a helpful tool for planning multiple crops per year, empowering community workers and volunteers, and keeping a record of what foods and soil management has happened over the years. The map is a simple visual tool that is a picture of the property with each zone of the garden differentiated—such as areas that are for habitat, corn, or water infiltration, or fruit orchard, and so on. Each plant type may require a different type of maintenance for the year for its successful growth, such as different watering amounts, soil management nutrients, pruning, or tying. The maintenance map illustrates where each plant type is grown for ease of communicating where to go and how to care for that specific item. One urban farm plot may have more than one map. Each map should be dated to indicate the months that it is used for. Seeing a series of maps will allow workers to see when plants will be harvested and replanted with the same plants, changed to another crop, or given a compost mulch top-dress. For example, if you are looking at a map that runs until August and see where the tomatoes are, you can look at the maintenance checklist timeline and see that they will be harvested at the end of August. Then you can check the maintenance checklist for the next food crop to be planted in that area. What is helpful is that overlapping areas and crops can be identified more easily with the mapping tool. Let's say you have tomatoes, peppers, and cilantro all sown next to each other for the summer. Then in the fall, you want to do an alfalfa cover crop over the entire area of the three crops that have been harvested. The fall map for the month after the harvest can illustrate the areas to be planted in alfalfa with any other details of the planting specifications that are needed for the soil, irrigation, or broadcast of the seed.

The Maintenance Manual

Setting the maintenance system criteria and goals into a written maintenance manual is a tool that will assist the management staff in overseeing the operations for the food landscape or urban farm. The manual also provides an opportunity to tie the operations structure to the annual budget and return on investment objectives, and provides a means to monitor the stakeholder goals. A manual records all of the data in one place so that it is easier to locate, share, and distribute to the appropriate people. Identifying the existing human capital to help establish and learn though the seasons about the ongoing maintenance should be part of the maintenance manual's operational tasks.

Another way to achieve learning objectives with maintenance efforts is to establish partnerships—like having a mentor such as a grandfather or other person from the neighborhood meet monthly with the community participants or hold a wisdom exchange to learn from each other about the weather, bugs, disease, water, seed saving, food harvest, and storage. Not only do these types of gatherings build community but they build shared wisdom and add to the ecoliteracy of the community.

A tricky part of maintaining a dynamic food landscape is knowing what to do with the changing conditions. A food system has many moving parts and considerations for those in charge of deciding on what to do to maintain the system. Each year, the soil, sun, water, and market for food, climate, and crop varieties can be different. For a new food system without the years of experience and observation of the nuances in that system, the decisions can be difficult. Each community likely has talent that knows the nuances of the ecology that your design will be built on. Rather than bringing in outside "experts," look for and engage the existing farmers and gardeners who have had the time and experience to deeply know the patterns of the local ecology. These folks can be the key to helping new farmers in the community successfully deal with changes and challenges they may have never seen before. Initiating learning partnerships and communities of practice with old timers and new farmers can help the growers of the urban food system continuously learn from each other, encourage resource and seed sharing, and build a mentoring structure that creates farmers for years to come. This can be as simple as a monthly potluck that rotates from garden to garden, house to house. You may only need a few people to identify that are a mix of new and seasoned farmers in the area who are willing to get it going and organize it for the first few years.

Soil and Organic Matter Management in a Lifecycle Approach

To ensure that the soil nutrients and the soil food web life are sustainably maintained, it is important to design a cycle for organic matter to be collected, composted, and reintroduced into the soil. This will ensure the correct organic matter percentage for soil life to thrive, and therefore the nutrients, soil texture, water, and plant health can be maintained most effectively. The waste material left over after harvesting food, trimming trees and plants, and cutting grass is valuable material that can be processed on site and reused. The design of the composting and processing space should accommodate the amount of greenwaste anticipated, and the rate in which the compost can be finished by whoever is in charge of it. Effective composting systems should have minimal smell and enough space to be turned frequently to speed up the process and make the material available to the garden faster. The balance between the amount of compost you will gain, the space and time required to make it, the labor needed to maintain it, and the costs that are offset by not having to haul in material should be calculated to determine the benefits and effectiveness of the system.

Consider the natural flow of the community through the space as well. Perhaps the participants in the food system will walk through the garden frequently if it is on the path to their bus stop, and would be able to drop off their food scraps from home regularly to add to the amount of compost for the gardens. People could drop their food scraps and wash out the container and leave it at the garden. The clean container might then double as a basket for fresh-picked foods on the way home. You might think about stacking other functions together as well. For example, the compost dropoff might be right next to the volunteer sign-up sheet and announcements, or next to the garden patches that need the most attention so they are easy to keep an eye on, and perhaps the herbs for cooking are on the path so folks can pick a bit for dinner on the way through. Mixing together routines with the design of the garden will allow for the most effective and efficient weaving together of the stakeholder's needs and daily routines to ensure that the food gardens are used and increase the quality of life for both. The gardens become a part of how people can get things off their to-do list of errands, and those errands can help the garden get what it needs—a reciprocity relationship that benefits both.

Identifying other opportunities for sourcing and processing organic matter in the surrounding neighborhood could prove to be beneficial to the soil management, without losing valuable growing space and aesthetics in the food growing system. The soil management program (Figure 5.15) could include collaborating with other landowners and communities to create compost that can be easily transported to the growing areas. Creating relationships and collaborations between plots of land and neighborhoods will help the overall effectiveness and energy efficiency of your design by reducing the need to ship waste material out and finished compost in for long distances. It will also be possible for food to travel one way and compost the other, thus creating exchanges of products that complement the other to satisfy the needs of the urban food system over a larger geography.

Program Schedule:

Schedules for Organic Transitions from Synthetics will vary depending on the initial condition of the soil, the budget available, along with client expectations and fertilizer used. Once soil organic matter is increased and biology is present, an organic maintenance program can concentrate on feeding the soil biology. A good system must have three components: habitat, biology, and food. Soils deficient in one of these will need more inputs to promote healthy soil building and related plant health.

For test sites and some instances, it will make sense to send in initial soil tests as well as followup tests to adjust the program. Remember to take pictures!

	Jan	Feb	Mar	April	May	June	July	Aug	Sept	Oct	Dec
Ideal Schedule											
Core Aerate										X	
Tea w/ Mycorrhizal and kelp										X	
Tea & Kelp			X							X	
Compost Top-Dressing										X	
Azomite										X	
Fertilizer App			X			X				X	
Budget Schedule											
Core Aerate			X								
Tea W/ Mycrorrhizal and Kelp			X								
Fertilizer App			X							X	
Fertigation Schedule											
Tank Filling			X			X				X**	

** If budget allows or soil conditions dictate, a third tank filling is recommended.

A soil assessment checklist (Figure 5.16) is a good tool to use anytime the soil needs to be tested or evaluated for its nutrient capabilities. The assessment checklist helps identify what is needed for the soil to be regenerated and maintained for optimal soil life and nutrients availability to the plant. Then the maintenance schedule can be created for planning the types of activities that are needed. It is recommended that the labor and equipment needs are listed if the management of the property is unaware of these needs. However, it is recommended that the assessment and planning be done by a very experienced urban organic farmer with good soil knowledge. The data collection form (Figure 5.17) is used by people who work on the property on a monthly or bimonthly basis to assess the changing and current conditions of the soil to determine if the maintenance plan is still correct and what adjustments might need to be made. It also allows workers with some soil knowledge to take assessment of the site and send the info to the expert for advice without expensive site visits. At the beginning of the project, the local experts can train others how to take an assessment.

Figure 5.15 A maintenance checklist for use in converting poor soils into healthier soils.

Soil Assessment Checklist

Date:

Name & Location of Site:

Manager:

Contact Info:

Circle the descriptions and answers that apply to the site's soil:

What type of plant material (circle one)? **Turf   Shrub   Ground cover   Color   Tree**

Did you send in a soil biology test? (Check the box)?   [] Yes   [] No

Did you *take* and *label* pictures of the *core* samples and the *plant* material?   [] Yes [] No

What type of soil is on the site? (Circle One)

Sand = Soil will not stay in a ball. Loose and single-grained with a gritty feeling when moistened

Loamy sand = A cast will form but it can't be handled without breaking and will not form into a ribbon. Soil feels slightly gritty.

Loam = A short ribbon can be formed but breaks when about ½ inch long

Clay loam = A ribbon can be formed. The ribbon is moderately strong until it breaks at about ¾ inch length. Soil feels slightly sticky.

Clay = The soil can easily be formed into a ribbon 1 inch or longer. Soil feels very sticky.

Organic Matter? Yes No

Mulch? Yes No   What kind of mulch? How deep is mulch?

Smell of soil = **Earthy   Rotten eggs   Good   Bad   No smell**

Irrigation DU is (Circle one): **Good OK Bad**

Type of irrigation (circle one): **Rotor Spray Drip Bubbler Netafim**

Is there a weed problem (Circle one)? **Yes   No** If yes, Name: _____   **Low Medium High**

How deep is the compaction? (How far can you easily probe the soil?)

   2"   4"   6"   8"   10"   12"   14+"

Are there any worms or bugs in the soil?   **Yes   No**

What color is the soil?   **Gray   Light brown   Dark brown   Sand color   Red**

Figure 5.16 An example of a soil assessment checklist for redressing soil needs for organic vitality.

Organic Data Collection Form

Name of Site	Today's Date

Circle What Is Being Used

Other? Describe:	Turf	Spray	Weeds
	Shrub	Drip	Pests
	Tree	Netafim	Disease
	Ground Cover		

List Products Used	Application Rate		
		per	
		per	
		per	

NOTES:

Notes - please BE SPECIFIC!

Use a 2nd page, if needed, to describe what is happening on site.

How is plant color, health, and growth?

What weeds, insects, or diseases are in the testing areas (list names)?

Customer reactions?

Notes of your experience (how to apply, smell, improvements, declines, etc):

Year:

Turf Areas *Equipment Used:*

	Jan	Feb	March	April	May	June	July	Aug	Sept	Oct	Nov	Dec
Compost Tea												
Organic Fertilizer												
Aerate												
Azomite												
Overseeding												
Compost Top Dress												
Labor per application												

Planted Areas (all non-turf areas) *Equipment Used:* *Products Used:*

Compost Tea												
Organic Fertilizer												
Compost Top Dress												
Mulch												
Labor per application												

Weed Control *Equipment Used:* *Products Used:*

Spot Spray												
Sheet Mulch												
Hand Pull												
Other (describe)												
Labor per application												

Soil Tests

Circle: Turf - Planted - Tree												

Figure 5.17 An example of a soil data collection form for use in collecting soil analysis information.

Riverpark Farm, Manhattan, New York City, New York

Riverpark Farm is a temporary portable farm (Figure 5.18) situated on the site of a stalled tower project in Midtown Manhattan. When construction of the West Tower of the Alexandria Center Campus was put on hiatus in 2008, the developer, Alexandria Real Estate Equities, Inc., decided to put the site to productive use (Figure 5.19). A partnership was formed between the Alexandria Center for Life Science and Riverpark, an adjacent restaurant, to develop much-needed green space for the neighborhood while also producing food to be used in the restaurant's menu. Chefs work closely with the farm, going through each day to see what is at its peak and ready to harvest and having that dictate the next day's menu.

Figure 5.18 Riverpark Farm's location on a stalled tower site creates activity and a productive landscape in a space that was vacant and unused.

Figure 5.19 The community chalkboard mural edges the urban farm.

Being a temporary installation, a little innovation was needed to allow for the flexibility required to literally pick up the site and relocate it in the future while also being an economically feasible solution (Figure 5.20). Riverpark Farm is constructed from 7,400 recycled milk crates, providing 3,200 cubic feet of soil on an area of 15,000 square feet (Figure 5.21). The milk crates provide an inexpensive modular system that can easily be put together, planted, broken down, and relocated (Figure 5.22). When construction of the West Tower resumes, the farm will be moved to another site of the four-acre campus.

Figure 5.20 The milk crates provide flexibility when there is a need for gathering or events.

Figure 5.21 The farm is located in a dense area of the city.

Figure 5.22 The farm can be easily picked up and moved to another location without much disturbance to the produce.

Two urban farmers manage Riverpark Farm, tending over 6,000 vegetable, herb, and flower plants and building compost from clean kitchen waste and coffee grounds during the year-long growing season. There are over 180 plant varietals growing, including carrots, cucumbers, basil, cilantro, greens, melons, tomatoes, and berries (Figure 5.23). In winter, after the autumn harvest, cold season–appropriate crops like spinach, beets, and carrots are planted, as well as a cover crop of winter rye to protect and maintain the soil until spring planting (Figure 5.24).

Figure 5.23 The farm grows a diversity of plants and sells them to the community.

Figure 5.24 The rows of milk crates allow for circulation paths.

Riverpark Farm has built connections within the community, opening its gates to neighbors and working with schools to promote *Plant-to-Plate* principles. During summer and fall, urban agriculture workshops are hosted on a variety of topics such as preparing beds, composting, and preparing soils for winter. A blog is maintained to relay helpful hints for the garden and the kitchen, translating what you are growing to what you are making for your meal and how to prepare it: www.riverparkfarm.com/_blog/Riverpark_Farm.

Design Team:

Client:	Riverpark Restaurant
Planter Design:	ORE Technology + Design

Understanding Harvest Distribution

When thinking about the future harvest distribution management, questions will arise depending on whether or not the landscape will be managed as a nonprofit or for-profit enterprise. Most nonprofit food landscapes are focused on meeting food desert and food security issues in underserved communities. These landscapes are typically based on community share and bartering models for harvest distribution with some produce sold at market stands. Others are developed as for sale market options, such as becoming part of a backyard or community supported agriculture model where members pay by the week or month subscription for seasonal fresh produce delivered to their homes or a nearby farm stand for pickup.

Harvest distribution management options include:

- For-sale options: farmers markets, CSAs, restaurants
- Donation options: food banks, charitable organizations, community organizations
- Volunteer community coop for personal use
- Community share and bartering options
- School lunch programs through gleaning programs
- Partnerships with organizations such as university or community college program
- Exclusive partnerships with local restaurants or food processing business such as bakeries, jam making, dried herbs and teas, etc.

Potential questions to consider for distribution when evaluating options are:

- Can connecting the urban ag landscape to the local community reduce the need for trucks in the maintenance of the landscape and the import of nutrients?
- How can you distribute food and gather resources at the same time?
- Can you make a small garden/landscape profitable by creating a supplement to an existing CSA infrastructure?
- Can designers/planners partner with farmers to find projects together by identifying land, communities, restaurants, and the need for fresh food to develop new markets and client bases?

- Can designers/planners partner with CSA farmers to develop landscapes that provide food and profits, and can be maintained by a local farmer for a new community?
- Since land aggregation leads to scale opportunities, larger farms are easier to produce enough food to turn a profit. Can multiple projects within a neighborhood or city pool resources into an aggregation entity?
- Is there an opportunity to partner with a local restaurant from the retail side that will also develop community-training programs for at-risk youths to learn culinary farm to table skills?

Tools and Storage Considerations

The amount of tools and storage required will depend on the size and scale of the urban ag landscape as well as the typology being developed. Tool sheds may need to function not just for storage but also for an outdoor kitchen, a supply cabinet, or a garden mangers office.

Understanding the uses for the shed will help to determine the size needed and the location in relationship to its use. A typical tool storage or tool shed for landscapes under 5,000 square feet can be about 80 to 144 square feet in size. Many modular sheds can be customized to fit the landscape's production needs and they come in sizes such as 8 feet by 10 feet or 12 feet by 10 feet, as an example. Storage sheds and containers can also be fun and whimsical and made from repurposed materials (Figure 5.25).

The use of greenhouses and hoop houses add to the seasonal reach for many climates. Greenhouses are also useful for seed and seedling propagation purposes (Figure 5.26).

To develop the quantity of tools and the appropriate-sized storage area for school gardens and community gardens, start by estimating the average number of volunteers, community members, staff, or students who will be in the garden at any one time. Another consideration in developing a tool list is

Figure 5.25 Tools and sheds can add whimsy and color to a garden.

Figure 5.26 Greenhouses help to extend the growing season and protect young seedlings and starts.

that tools can be used through cooperative ownership, thus keeping the budget to a moderate level. For some urban ag food production tasks, it is customary to ask volunteers to bring some of their own personal tools to use, depending on how big the garden or what the landscape workday will be. Generally, there is a need for smaller tools for planting and digging beds than for fewer larger tools. Numbers of tools would be related to the number of people that you can foresee doing any one task at a given time.

Items to have more of:
- Trowels
- Gloves
- Spades
- Digging forks

Seasonal tools
- Hoop houses
- Bamboo stakes for vertical plants
- Twine
- Shade and bird fabric

Items to have fewer of:
- Large brooms
- Rakes

- Wheelbarrows
- Shovels
- Pruning saw
- Hand pruners
- Hoses
- Heavy machinery—can be rented

Adapting the Maintenance Framework to Smaller Projects

Smaller projects such as personal food landscapes including home gardens and backyards (also called edible estates) or small window boxes and planters for a restaurant do not need a detailed maintenance plan, but they do require a simple job description and the use of a daily garden journal to keep track of the garden tasks, seasonal harvest plans, and simple checklists of what has been done.

Trial and error is also a useful hands-on way to fine tune a maintenance plan in small food landscapes. In the case of the Bar Agricole restaurant in San Francisco, the three herb planters in the outdoor dining terrace are maintained by the sommelier, who has a passion for food landscapes and the farm to table philosophy of the restaurant. Having taken classes at Alemany Farms, he also writes a food blog and works with the chef to make sure the seasonal variety is accounted for. Through trial and error, he has learned what is best to both look at and smell while dining and how much is needed to grow for different herbs to utilize in dishes or cocktails. The simpler the process can be to ensure good results, the better.

Labor Considerations and Green Jobs

Labor considerations for many urban ag landscapes typically rely on a combination of paid staff members supplemented by a larger numbr of community volunteers. Labor is also a reflection of the urban ag typology and varies depending on the management structure and business structure. There are do-it-yourself scenarios, community volunteers, paid professionals such as garden coordinators, farmers and master gardeners, support staff, and management supervisors. In the case of some landscapes, it is all done by community volunteers. Even for-profit models rely on volunteers to help keep costs down and create synergies with local community groups for education and stewardship opportunities.

For larger urban ag landscapes, there is the potential to create the opportunity for new green jobs centered on education and production for educators, mentors, garden coordinators, urban farmers, harvest delivery services, school lunch harvesting and gleaning programs, website designer, specialty installation services such as food landscape roof planters or hydroponic systems. Programs such as Green Jobs For All, founded by Van Jones, focus on the green economy and support the creation for urban ag jobs. A question to ask is whether there are green-collar job programs in the community that could support a mentoring program—if so, tap into that resource.

Organizing volunteers is a big task. Some organizations use Google Documents as a way to keep volunteer tasks and schedules coordinated without having to send multiple e-mails and reminders. Some organizations use a general list serv sent electronically to let members know about upcoming needs or events. Flyers are also a useful tool to notify the community about special events or programs that might be of interest. Having a website is of particular use for outreach to the general public about volunteer opportunities, educational programs, and events. More information on websites and marketing is covered in Chapter 6.

VF Outdoors Corporate Campus, Alameda, California

VF Outdoors, the outdoor division of VF Apparel, constructed a new corporate headquarters campus in the Harbor Bay business park in Alameda, California. The campus is located on a 15-acre parcel near the Oakland Airport and is adjacent to the Alameda Bay waterfront. The campus plan includes four two-story office buildings of 205,000 square feet with a cafe, employee fitness center, 35,000-square-foot outdoor courtyard with floating meeting decks (Figure 5.27), outdoor yoga court, and conference facilities.

Figure 5.27 The edible garden further enhances the campus' sustainable agenda, which also includes the nearby native habitat butterfly gardens and solar parking lot panels.

The site plan also includes an organic food garden (Figure 5.28), a butterfly garden, a future climbing wall and outdoor sport court, and waterfront trails.

The challenges for the landscape architect team, April Philips Design Works, led by April Philips, included stakeholder buy-in of a completely native landscape to create habitat gardens, though the company had a vision for a highly sustainable campus. The edible garden started out as a smaller garden of 3,000 square feet, but during construction the CEO realized it needed to be larger and more related to the wellness cafeteria, so the garden was redesigned to be 10,000 square feet.

The design idea is for the creation of a premiere organic garden in an office campus environment where elegance meets pared-down essentials (Figure 5.29). In addition to providing year-round produce for use in the wellness cafeteria, the landscape is designed to be used by the facility's employees, and to encourage gathering in the garden for socializing and relaxation. Toward this end, the steel raised beds help define gathering areas—one with movable seating beneath an ancient olive tree (Figure 5.30) and a central space with a communal picnic table and built-in tabletop planter.

The garden's perimeter vegetation brings the adjacent native meadow and butterfly garden plantings inside the fence, helping the space to feel integrated with its surrounding context. An ancient California Mission olive, sourced locally, anchors the garden and can be harvested for olives semi-annually while providing a great place for informal gathering under its canopy. Similarly, the garden's materials help to visually integrate it with the larger landscape. A sustainably harvested wooden fence helps to keep out larger wildlife predators and provide a buffer from the parking lot, as well as providing a canvas for espaliered fruit trees and a living herb wall of woolly pockets. The fence also forms a semi-secluded compost and tool storage area, keeping these elements out of view but easily accessible.

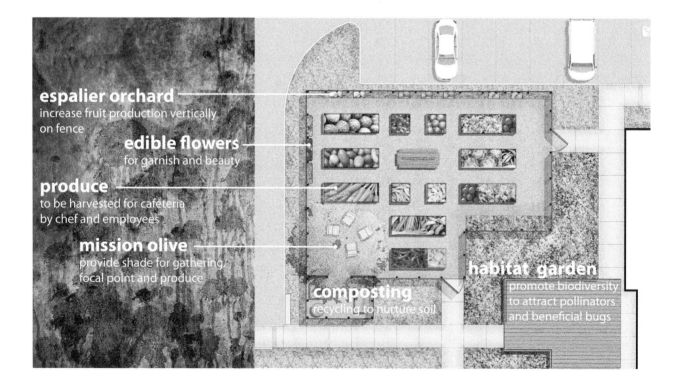

Figure 5.28 The 10,000-square-foot edible garden was originally only 3,000 square feet. It is anchored by a California Mission olive tree that can be harvested twice a year. The other main components are the espalier orchard, the raised steel planters, the edible flower wall, the composting area, and the adjacent habitat garden that supplies pollinators.

Figure 5.31 The garden is harvested by the VF garden team for the cafeteria and for personal use. Excess produce is donated to local food banks. The solar panels in the parking lot add to the synergy of the sustainable goals.

Figure 5.32 The raised beds are compleme grown in vertical Woolly Pockets on one ed trees on the other.

Currently, April is working with the client's garden group and the chef group to set up the seasonal garden plans and the volunteer resources. Issues have included integrated pest management of the multitude of critters on site and salt tolerance of the vegetation because of the bay area location. The project is seeking LEED Gold campus certification for the project and certification for the Bay Friendly Landscape rating (Figures 5.31 and 5.32).

Design Team:	
Client:	VF Out
Project Manager:	SRMern Group
Landscape Architect:	April Ph
Landscape Contractor:	Cagwin
General Contractor:	JM O'N
Size:	15 Acres Garden
Year Constructed:	Autumn

assist young people ages 16–22 from severely at-risk communities who desire to make a positive change in their lives. Their students typically are facing a vast array of challenges, from extreme poverty and high school attrition to homelessness, violence, and participation in the juvenile justice system. However, they all have a strong desire to break that cycle and become productive, contributing members of society. Café Reconcile is a nonprofit high-end food experience providing top-notch service and exquisite dishes on the menu. The staff consists of expert restaurateurs, chefs, and urban farmers with a passion about growing local food and running great restaurants. Young people of disadvantaged communities can apply to become an apprentice of the Café to learn everything they need for a career in the many high-end restaurants of New Orleans. Café Reconcile focuses on locally grown organic foods, and connects the apprentices with first-hand experience with how urban farming works. The food grown in their urban plots ends up in the kitchen, where the education continues on what foods work best together to create dishes that stand up to New Orleans dining expectations (Figure 5.34). They also learn the economics of the restaurant and how viable an option it is to purchase locally grown foods versus from big distributors. Finally, the food is served in high style and with every detail of the dining experience being considered and learned. The food scraps then are composted for the urban gardens the food and the apprentice started from to complete the cycle. This model is a great example of how an urban food system can create many opportunities for people to improve the quality of life for themselves and the community connected to it (Figure 5.35).

Figure 5.34 The urban kitchen garden is a few doors down from the restaurant.

Mentoring and Longevity Models for Operations

There are a number of mentoring and longevity models to consider when developing educational and stewardship programs around urban agriculture. A brief summary of possible methods follows; we will take a look at some of these examples in more detail.

Educational and Community Engagement Methods

Community engagement methods include community workshops on resilient food ecologies, designing for food systems, budget and return on investment math, urban agriculture practices curriculum, education and stewardship, keeping the volunteer workforce energized, applications for urban ag technologies, and ongoing recognition programs for volunteer contributions, for example.

Partnering with a university or other type of research or educational facility is another way to weave the community into the project in order to gain a sense of ownership and buy-in from the community. Courses could be created with students and existing professors that ties into their current curriculum but also cater to the local community needs.

Connecting with Restaurants

Are there restaurants that may want to co-own the property or prepay for the right to the food before other consumers? Can a relationship with a restaurant ensure that the edible produce will be purchased, therefore reducing the risk to invest in labor, materials, and size of garden to ensure a profit? Restaurants offer a way to also create a link between food growing and food preparation though a culinary tie to the landscape. This method can offer farm-to-table skills and restaurant business skills to the local community.

Café Reconcile New Orleans is a community of concerned people committed to addressing the system of generational poverty, violence, and neglect in the New Orleans area (Figure 5.33). They have an innovative life skills and job training program to

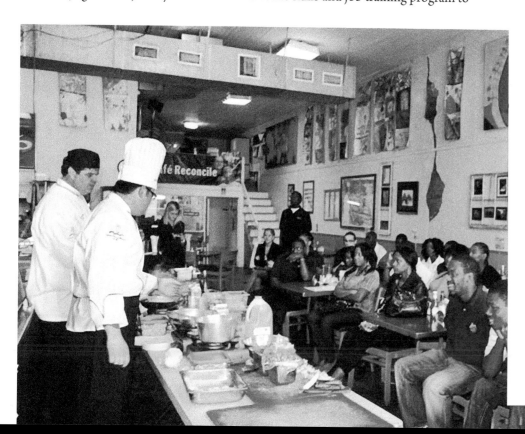

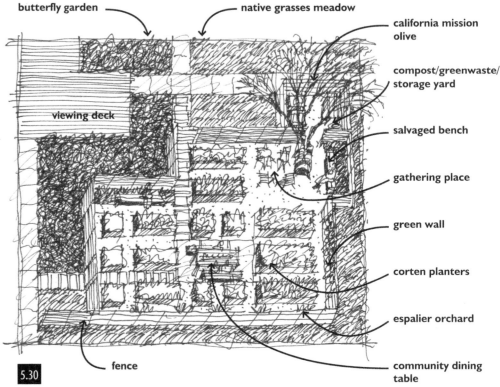

Figure 5.29 The custom, locally made corten steel planters leave more space for planting and circulation, and create a clean, modern look to match the rest of the campus.

Figure 5.30 The original sketch of the heritage olive was one of the key ingredients that got VF and its chef excited about the campus food garden.

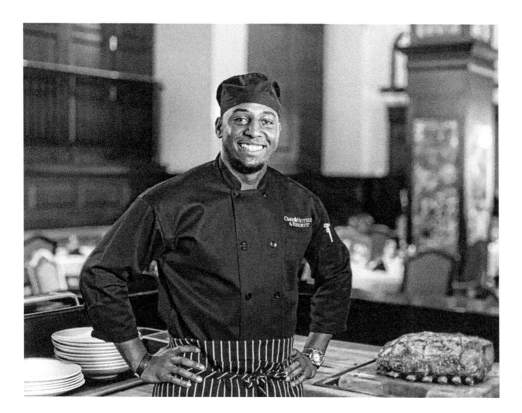

Figure 5.35 A graduate of the Café Reconcile job training program.

Training and Mentoring

Most contractors will need training for large urban ag projects, and should consider the need to partner with local farmers, nonprofits, and urban farmers and to subcontract with a mentor to work with the crews for 12 months in order to learn the four seasons of maintenance in the gardens. Questions stakeholders need to consider in thinking about community interdependence include the following: What organization can your food landscape support? Can you figure out how your design will create healing in the community? Would a green job training program be of benefit to the farm and the community?

City Slicker Farms in Oakland, California, has a mentoring program that ensures people will be able to grow their own food. Mentors with farming skills are paid for their knowledge, and people who become passionate about urban farming can find pathways to do more and become mentors themselves. They help give families small gardens in their yards and match them with an experienced gardener that will check in to help with the garden monthly and as needed. This helps the new gardener gain the confidence needed to try it out, and helps ensure that the investment in the garden is not wasted. As the year progresses, the mentor teaches what can be grown and the skills needed for each month of the year so the farm can be productive. The mentors and garden recipients are chosen through an application process that helps manage the quality of the experience for all involved. City Slicker Farms also has large plots of land throughout West Oakland that provide food for markets, but also a training ground for volunteers and students to learn urban farming techniques for getting started or expanding on their small gardens. The larger plots also create an opportunity for

Figure 5.36 Workshops and community outreach events add to the trust building within the community and are a strong link to the training programs.

community members to learn together and share what they are learning in their own yards. The habit of sharing with each other and learning together is cultivated through having days set aside for intentional gathering on a consistent basis so people can count on it happening (Figure 5.36). The consistency is also a good way to build trust and comfort in getting the support needed from a structure of managing logistics. It makes it easier to build the habit of showing up and sharing as an important thing to do in life.

A mentor program for younger kids is in New Orleans, called Market Farmers. It was founded by Jenga in the Lower Ninth Ward. Design projects include allowing for community participation at key moments in the design, installation, and maintenance of the gardens in order to ensure that the community has decided to continue investing their energy and time, and validate that they are able to follow-through with the long-term care of the garden. It focuses the education on developing skills that would allow the students to become part of the fabric of the community.

Some restaurants want to invest in owning land or working with an urban farm. Farms on commercial or residential community property, such as Sand Hill Farms at Prairie Crossing in Illinois, offer a farm training investment program for new farmers where they provide a plot to farm for five years, along with resources to learn sustainable farming practices before the farmers are ready to go off and develop a farm landscape on their own.

School Edible Garden Programs

There has been an explosion of edible school gardens and educational programs in the past few years. Bringing school kids into the gardens helps the project to have value for generations. These stewardship programs build the skills and ecoliteracy of the next generation so we don't have to work so hard to educate our communities on how and why to create urban living spaces that include food. The Center for Ecoliteracy has educational programs and resources on their web site that any educator can download for free. *From Asphalt to Ecosystems* by Sharon Danks and *How to Grow a School Garden* by Arden Bucklin-Sporer and Rachel Kathleen Pringle are two great resource books for aiding the development of school gardens.

Master Gardener Programs

Master gardener programs are expanding into sustainable and edible educational programs all over the country. They usually need places that are new and innovative to teach classes, to volunteer, and to connect with the community. They can help manage the ongoing education of the gardeners and managers if there is turnover.

Job Training Opportunities for Colleges

Urban food landscapes offer job-training opportunities for the new food economy, including farming, marketing, sales, business development, ecological restoration, and environmental studies education with local universities. Design courses with local architectural programs are another route to explore especially as it relates to designing for food systems, ecological health and community health.

Multigenerational Connections

Active seniors have skills that need to be passed down to the next generation to preserve cultural farming techniques. For urban agriculture to be effective, families need togetherness and a sense of self-worth. Design the landscape for everyone's needs to make it a natural place for everyone to enjoy, and each generation may suggest going there.

Company Gardens That Give Back

Some visionary companies believe in creating a workplace that is a sustainable lifestyle environment that promotes environmental stewardship and gives back to the community. These companies such as VF Outdoors are providing employees perks such as fresh organic food for the cafeteria grown from their own garden, giving philanthropic donations to local food banks or charitable organizations, and creating outdoor places that are restful and peaceful for the employees benefit. This trend may be indicative of a transformation into a healthier, more sustainable lifestyle and not just a trend. Ecoliteracy and building community are at the heart of this type of program.

Sacred Heart Preparatory Organic Vegetable Garden, Atherton, California

Sacred Heart Preparatory's organic vegetable garden (Figure 5.37) is part of the Michael J. Homer Science and Student Life Center. The Michael J Homer Science and Life Center was LEED Platinum certified. The organic garden is just one of the outdoor learning classrooms that surround the building (Figure 5.38). The 1,000-square-foot (135 feet long x 7.5 feet wide) organic vegetable garden generated so much interest that Sacred Heart converted an additional 9,000 square feet of campus ornamental landscapes into organic vegetable gardens. In addition to the school's enthusiastic garden champion Dr. Stewart Slafter, the designers, SWA Group, contributed to the garden's success by orchestrating the garden to flank the dining terrace (Figure 5.39),

Figure 5.37 The linear edible garden at Sacred Heart.

Figure 5.38 The dining terrace during lunchtime, where ingredients from the garden are used when available.

Lifecycle Operations 219

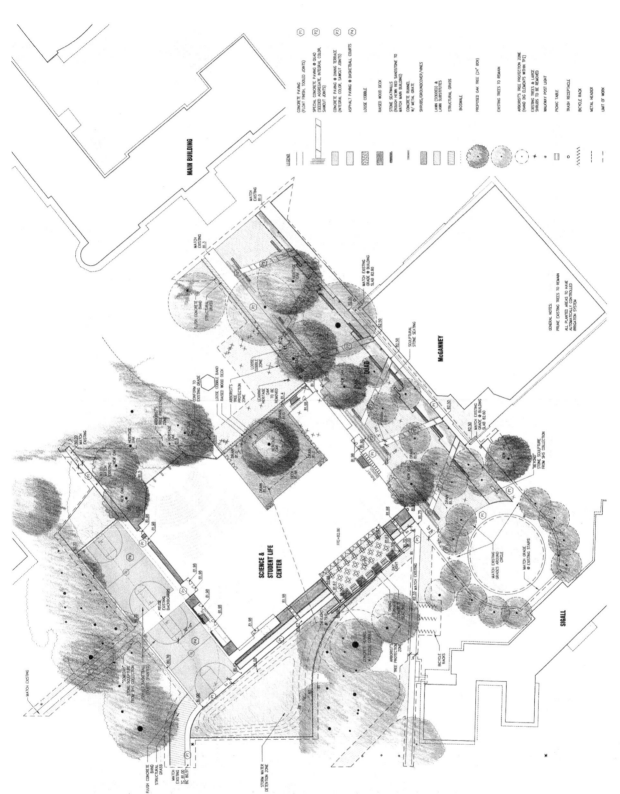

Figure 5.39 The site plan of the Homer Center building includes a number of outdoor classrooms, of which the edible garden is one.

Figure 5.40 The straight walkway is flanked by rows of vegetables.

on the south side of the building, and providing a clear, simple structure—a straight flat walk in front (Figure 5.40), many short, perpendicular planting rows, each with only one type of vegetable, and a "Little Ollie" hedge behind it so that the garden would look tidy in all phases. Water comes from the campus well.

Challenges for the design team included overcoming skepticism that the garden would be tended for more than the initial burst, and that it would look acceptable, but within months the additional 9,000-square-foot garden was started because of the project's success. There are a number of synergies between the two gardens, such as waste and water management and production management.

As part of additional overall sustainable strategies, decades of accreted paving were tidied up to create "designer parking" along the campus's 1000-foot-long row of 50 heritage olive trees estimated to be more than 100 years old, which Sacred Heart harvests and presses. The school community has come together in early November to harvest the olives from the trees since 2009. The fruit is then pressed into a premium olive oil, bottled, and sold as a fundraiser. The education of learning how to manage the trees, produce a high-quality olive oil, discover what makes an olive oil extra virgin, and what factors affect its taste are all part of the lessons learned for both students and adults to appreciate the entire process from tree to table.

Under the supervision of Dr. Slafter, the 10,000- square-foot organic garden is maintained and operated by Sacred Heart students and faculty year around as part of its Environmental Science and Global Studies courses. The school is the first in San Mateo County, California, to be approved by the Department of Environmental Health Services Division to use fruits and vegetables grown in its own organic garden for preparation in its school cafeteria. In addition to feeding faculty, staff, and students, food grown and harvested in the organic garden is donated to local charities like St. Anthony of Padua's Dining Room in Redwood City, California, and Sandwiches on Sundays in Menlo Park, California. Once a month, faculty and students harvest and prepare food to be served to the homeless in San Francisco's Civic Center as part of a program called "Food, Not Bombs."

Slafter said, "One of the chief purposes of our gardens is to teach students how to be stewards of the Earth's resources. We're teaching young people about sustainability and how they can grow food and know about every step of the process so they understand that agriculture can be continued on a permanent basis."

Design Team:

Client:	Sacred Heart Schools
Landscape Architect:	SWA Group
Architect:	Leddy Maytum Stacy Architects
Garden Leader:	Dr. Stewart Slafter, Sacred Heart Schools

The Finance Plan

In Chapter 3 we discussed that it is important to develop a preliminary business plan early in the design process. It starts out as a draft outline identifying the key lifecycle considerations of the project including the budget drivers—both incoming and outgoing, selecting the business model that best fits the project, outlining the future operations structure, the stakeholder decision making process, the future marketing and outreach potential, and the green job potential. Without understanding the project's potential overall budget parameters, a project will not have a clear enough roadmap to tackle its own growth and development.

The finance plan takes the information from the preliminary budget and refines the numbers based on the project's parameters as they have evolved during the design system integration phase. The proposed business model that begins as a lifecycle diagram to promote discussion with the stakeholders is now refined into a working budget that identifies operations, return on investments, labor considerations, and public and private funding sources. There will be a major difference between public and private developments when it comes to public and private funding sources.

Finance Challenges

Funding can come from a combination of sources such as grants, fundraising, private–public partnerships, corporate benefits package, institutional donations, and more. Using the resources available, refine the list to the top-ten sources and potential partnerships and begin to evaluate the effort required to follow through on them to make the yes and no decisions in moving forward:

- If the preliminary budget was set up previously, start with that to update the number to the most recent design decisions. If it has not yet been created, now is the time to do it. Refinements of the numbers will be based on the updated plans and decisions. New ideas or partnerships may have been found that can be factored into the budget at this time.

- Using the communications protocol that is in place with the stakeholders, allow for a feedback loop with the draft finance plan.
- Examine if the proposed business model established early on is still the right and feasible choice to meet the visions and goals. Is this model the one that best fits the needs of the community so that it will thrive into the future?
- Identify the community's existing food shed opportunities and constraints that the project can tie into and determine if the budget is addressing these connections. Is there a cost to providing the missing links? If so, record that into the budget.
- Identify the budget costs of the proposed maintenance and management system. Is there a management system in place? How will it be operated? These questions should be answered at this time before the project is completed.
- Reexamine the return on investment numbers to see if they still add up. Are there ways to increase the return on investment that have not yet been identified? Is scale investment a potential solution to pursue?
- Set a budget number for the marketing, education, and outreach goals. Has the project been able to consider how to incentivize it for benefiting the community and local ecology? Have partnerships with organizations been identified for stewardship, education and green job training programs? Have a discussion on green job training, education programs, community and business partnerships, school and institutional partnerships, and mentoring and longevity models to determine next steps for those programs and how they tie into the operations budget.

Resource Guide for Funding Sources

There are a number of avenues to pursue funding for urban ag landscapes. This includes a combination of grants, private donations, philanthropic donations, organization partnerships, and others. Funding sources will depend on the urban ag typology and the vision and mission of the food landscape. The better the vision is defined and represents a broad spectrum of benefits for ecology, culture, and economy, the better the chance of finding a funding source that might fit your need. Here is a sampling of potential sources to look into:

- ***Funding partner organizations:*** Look for funding guides available that are based on the specific typologies or prototypes your project might fall into. Think about searching for grant mechanisms that provide for starting budgets and ongoing programs. Research locally for organizations that have a track record for supporting communities and programs that fit the urban ag vision. Issue topics for funding generally include: nutrition, agriculture, education, obesity, and community health.

- *National foundations making grants for food systems:* Foundations tend toward creating systems that only need money for a few years to get off the ground.

- *Public and private funding:* Seek out research and grant funding for programs and continued maintenance for the garden. Consider applying for the grants with an existing nonprofit in the area in order to ensure community ownership of the project.

- *Partnering with cities to measure ecosystem service:* Perhaps there is an opportunity to partner with cities to measure the ecosystem services that the landscapes are providing to promote further investment in edible landscapes throughout the city. Look to the local government agencies to find champions for local food economic development, green jobs programs, congress people hoping to write legislation, and nonprofits lobbying. Try to connect the garden to their efforts to encourage the change they are trying to make while potentially helping your garden become funded and more connected to the community—marketing through social change for the good.

- *Partnering with municipalities:* It costs municipalities a lot to maintain clean air and water, manage watersheds from runoff, restore creek systems to improve fisheries, address medical costs from toxins, undertake brownfield remediation, clean up Superfund sites, and so on. If you can measure the benefits of an edible landscape that also uses sustainable maintenance techniques that provide solutions to those maintenance issues, then the city might allocate money to ecological land management through sustainable landscape architecture and contracting.

- *Tie into organizations or enterprises that are commercial or educational:* We discussed the food connection with restaurants earlier in this chapter as a means for partnering potential. Other enterprises include schools, universities, or community colleges that may be able to provide an academic relationship that supports the garden or farm. Local businesses also may want to give back or offer an employee good works program. Others may have land available they are willing to donate for the project. There are many opportunities to find relationships if you look for them.

- *Websites for grants:* doing an internet search is a good way to start looking for funding opportunities that might be available. Rebel Tomato is one that provides information on how to look for ones to fit your needs. The ehow web site is also a good place to start for ideas on how to search. The EPA has grants for starting community gardens in brownfields. Contact your local United States Department of Agriculture or Farm Service Agency office to see what funding is available. The Sustainable Agriculture Research and Education, or SARE, program provides grants for agricultural projects led by youth 8 to 18 years of age. Cities also have various programs to research. For example, the HPNAP seed grants by the United Way in New York City is the Hunger Prevention and Nutrition Assistance Program that supports urban agriculture.

Slow Food Nation Victory Garden, San Francisco, California

Figure 5.41 The temporary victory garden was part of the Slow Food nation weeklong event in San Francisco.

A temporary installation, the construction of the Slow Food Nation Victory Garden began in the summer of 2008 with volunteer urban gardeners removing 10,000 square feet of turf from the plaza in front of the San Francisco City Hall. Volunteers used simple materials like burlap sand bags (Figure 5.41) and donated soil to construct beds for the temporary food-producing garden. The victory garden was finished on July 12, when 250 volunteers planted thousands of seedlings for the garden's heritage organic vegetables. The garden was built in collaboration with the Victory Garden 08+ Program and the Slow Food Nation to highlight food justice and urban agriculture issues.

The Slow Food Nation Victory Garden demonstrated the potential for local food production within the city limits and provided fresh, healthy food for those in need. It was constructed on the site of a post WWII Victory Garden at the San Francisco City Hall and Civic Center and highlights a lost urban agriculture tradition in San Francisco: during WWII the City produced 30 percent of its food in urban victory gardens. The garden was able to harvest approximately 100 pounds of fresh organic produce and donate it to the San Francisco Food Bank to distribute throughout the City. Located in the Civic Center public park and adjacent to the Tenderloin, a neighborhood known for homelessness and crime, the garden was an effort to rethink what is generally considered appropriate for urban open spaces and employ the space as a productive landscape.

A strong community of organizations and volunteers collaborated to promote the success of the Slow Food Nation Victory Garden. City Slickers of Oakland and Ploughshares Nursery of Alameda propagated the seeds donated by Seeds of Change. Pollinator attractor plants were provided by Coevolution Institute. Experienced and novice volunteers tended the crops and provided maintenance for the project.

Initially scheduled to end on September 21, Mayor Gavin Newsom (Figure 5.42) requested that the Victory Garden stay in place through Thanksgiving 2008. The garden began dismantling on November 23 and materials were donated to Project Homeless Connect, a local organization serving San Francisco's homeless population with skill-training.

Figure 5.42 The garden was planted and maintained by several community organizations and over 250 volunteers.

Design Team:

Landscape Architect: Rebar Art & Design Studio

Organization Team:

Victory Garden 08+ Program, Slow Food Nation

Resource Team:

City Slickers Farm, Ploughshares Nursery, Co-Evolution Institute, Seeds of Change

Resources

Bucklin-Sporer, Arden, and Rachel Kathleen Pringle. *How to Grow a School Garden: A Complete Guide for Parents and Teachers*. Portland: Timber Press, Inc., 2010.

Danks, Sharon Gamson. *Asphalt to Ecosystems: Design Ideas for Schoolyard Transformation*. Oakland: New Village Press, 2010.

The children's garden at Lafayette Greens in downtown Detroit.

CHAPTER **6**
Outreach and Community

Atlanta Botanical Garden
Atlanta, Georgia

The Atlanta Botanical Garden is home to a wide variety of gardens, each showcasing unique plants and unique design for the past 35 years. Their mission is to develop and maintain plant collections for display, education, research, conservation, and enjoyment. They have enjoyed being a must-see destination place for many years, but it was the 2010 openings of the Canopy Walk, Cascades Garden, and Edible Garden (Figure 6.1) that solidified ABG as a premier destination spot for the young, old, and urban hipster alike. The Edible Garden is perhaps the most memorable and certainly the most delicious. Home to not only a wide diversity of plants, but also a variety of unique spaces designed to engage, educate, delight, and even taste. The garden demonstrates to visitors that edible plants can be beautiful as well as functional (Figure 6.2). Most visitors go home excited to cook a meal with the plants they have just learned about in their garden experience.

Figure 6.1 The Edible Garden aims to demonstrate the aesthetic value as well as the horticulture value of edible plants.

Figure 6.2 Colors and textures were exploited to show the design capability of edibles in architectural spaces.

In the 2010 master plan expansion projects at ABG, the focus was placed heavily on ecosystem sustainability, conservation, and driving visitation to the garden. Mary Pat Matheson, executive director of ABG and one of the visionaries behind the recent expansion noted, "We don't have what zoos have—cute warm and fuzzy pandas—so an edible garden was a chance to have 'fun with the Fuzzy.'" Using the space of a former parking lot, the garden was championed with the slogan "from asphalt to asparagus." According to Mary Pat, the Edible Garden was seen as a place to forge connections with people and plants since everyone needs to eat and the link between food and health was a perfect one to explore as an educational opportunity. Tying all of this into Atlanta's foodie culture was a cinch.

Inspired by otherworldly entities like crop circles and UFOs, the Edible Garden's site plan (Figure 6.3) makes elegant use of geometry to define areas for row crops, raised beds, a pool for aquatic crops, and espaliered fruit trees. The goal of the garden is to first engage visitors and draw them in, and then to educate them.

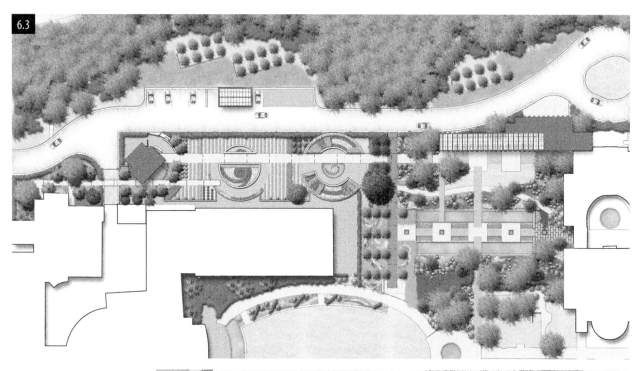

Figure 6.3 The geometric plan highlights the beautiful and minimizes the "uglies" as you stroll through the garden.

Figure 6.4 Herbs are particularly well suited for the fragrant living wall.

Figure 6.5 Bold swaths of contrasting colors highlight the design possibilities and Chilluly glass lilies highlight the composition.

Figure 6.6 The raised beds put color and texture at eye level for easy viewing.

Toward this end, a 50-foot-long vertical herb wall (Figure 6.4) provides a fragrant, visually striking, and texturally stimulating element as the springboard into the edible garden. This wall was inspired by a trip to see Patrick Blanc's vertical wall at Musee Brownlee in Paris. Similarly, a terraced vegetable garden amphitheater puts plants right at head height, allowing guests to look into the plants instead of down on them from above (Figure 6.5). Functionally this technique also helps to put what Mary Pat calls the "uglies" in the background rather than front and center.

An important part of the seasonal plant selection has been to understand which plants provide beauty and function and which plants are less aesthetic and thus should be put in less visually challenging locations. It was important to identify the list of functional and beautiful plants up front as well as the uglies to maintain the garden as a beautiful experience for visitors to return to again and again. Along with horticulturist Colleen Golden, some of the beautiful reliable plants identified include sweet potato vine, basil, and fennel, while tomatoes are typically tucked away in the back rows since they are messy. Plants that have surprised them have been okra, which has turned out to be quite stunning, as has the combination of beets with tulips. Mary Pat's philosophy is to start simple to address what she calls "plant blindness" by focusing on the design qualities of the plants and presenting them in a visually bold manner so that visitors will see the plants in new ways (Figure 6.6).

Her theory is that once you have seen them, you will think about them and perhaps try new things. Art is incorporated in unique ways in the garden as well during the year to add to this type of seeing.

The garden's outdoor Kitchen Pavilion (Figure 6.7) provides the culinary experience of the Edible Garden, although there are places along the circulation paths where visitors are invited to sniff and touch. The Kitchen Pavilion hosts a variety of culinary classes and demonstrations, using plants from the garden, including the Garden Chef series, which brings Atlanta's top chefs to the kitchen. The pavilion can be rented for private parties, and visitors can even come once a week to enjoy a cocktail during an evening Garden Chef class (Figure 6.8).

Figure 6.7 Culinary classes have become a popular draw in the garden's Kitchen Pavilion.

Figure 6.8 The space has comfortable gathering areas drawing together food and celebration.

Figure 6.9 The garden employs edible plants of all sizes and colors to build depth.

The garden publishes a list of the chefs' recipes using foods grown in the garden. Programs and classes are ongoing and popular at ABG. Children's summer cooking camps typically sell out in less than 8 hours. After the Children's Garden, the Edible Garden is ABG's second most popular outdoor garden. It is the combination of environmental stewardship, playful discovery, and unique horticultural beauty that has made the Edible Garden one of Atlanta's highly regarded destinations (Figure 6.9).

Design Team:	
Landscape Architects:	MESA Design Group, Tres Fromme and Studio Outside
Collaborator:	AECOM
Architect:	Axios Architecture
Client:	Atlanta Botanical Garden Mary Pat Matheson (Executive Director), Colleen Golden (Horticulturist)

232 Designing Urban Agriculture

The Outreach Sphere

The outreach sphere is the lifecycle link to the design and planning process of urban agriculture landscapes. Designing and constructing these landscapes is perhaps the easiest part to do in the lifecycle process. The outreach sphere (Figure 6.10) consists of funding, marketing, education, stewardship, research, policy, and advocacy.

More specifically, these categories translate into developing the project's marketing concept and maintaining the brand; expanding outreach from grassroots to mainstream education; developing an annual funding budget program; developing programs that

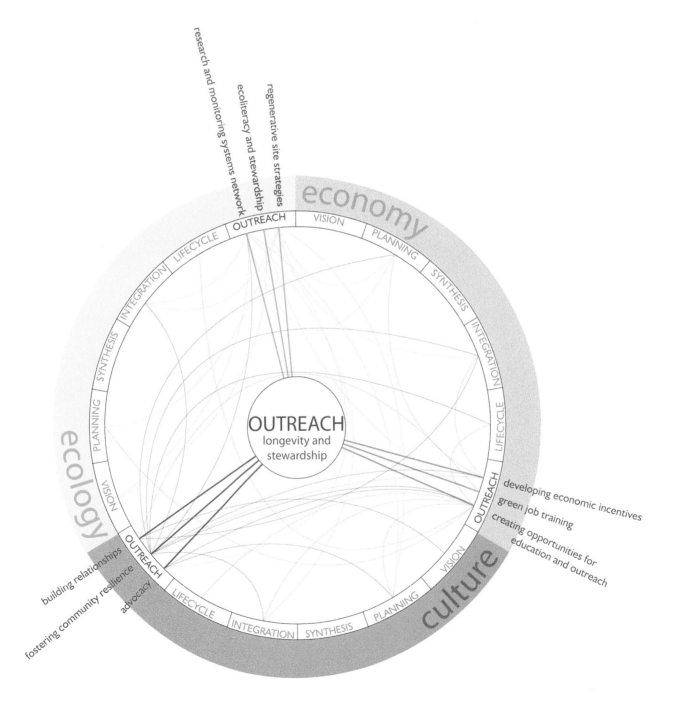

Figure 6.10 The outreach sphere diagram.

foster green job creation; developing partnerships that add complexity to the outreach and breadth of community building; establishing an operating system for monitoring and research that can be used for case study data and funding metrics, resource protection and advocacy development; creating policy and zoning in support of urban agriculture; and lastly, developing a communications format that links all of these items with the community stakeholders and partners for feedback and decision making.

The business plan is the main tool that can be used for achieving these outreach outcomes. Integral to the design process is setting up the framework plans in the business plan development for the immediate funding, marketing, and educational outreach of the project when the project is ready to be accessed or open to the community. Although the stakeholder team might change from the design phase to the ongoing operations phase, in general there will most likely be stakeholders who will remain or who will transition as new stakeholders come on board. Maintaining the vision is the key to a successful outreach plan leadership.

The outreach plan will vary with the urban ag typology and the vision and goals of the project. Components within the plan could include a plan for integrating the research and monitoring aspects of the project or it could also consider how the project might be affected by or would address local policy and advocacy outcomes that could impact the community in positive ways. The first thing to understand is what the outcomes are that you wish to reach through the outreach plan.

Funding and Grants

Both public and private urban agricultural landscapes will require an annual budget to operate even if they are to be maintained primarily by volunteers. Understanding the annual budget is the first step in looking for continuing financial support. Developing the annual funding mechanisms that allow for managing and operating them on an ongoing basis can be the more challenging part of the equation. The business plan is the mechanism to identify the funding opportunities for the first few years and the steps to achieve them. Funding options come from many sources, such as donations, grants, market sales, fundraiser events, CSA dues, other types of money and barter-based systems, and government funding. USDA funding includes programs such as the Peoples Garden and Know your Farmer, Know your Food. It makes Farm to Food Grants available as well.

Grants are also a solution and vary by project type and size. Local community grants, federal educational grants for schools, and nonprofit government grants are all potential funding opportunities. Grants are also provided by philanthropic organizations set up by a corporation entity such as Honda's sustainability grants, or Whole Food Markets school garden grants. These vary in size in scope from as low as $250 to as high as $25,000. Check your local businesses, as well as libraries for these resource lists.

Some cities and counties offer local community grants for proposals that are aimed at serving the community in a way that promotes the values and initiatives of the area. If you keep up with your community hot issues, this is sometimes a useful avenue to take via local politicians and supervisors. Participate with food advocacy issues and network with other community members who have similar interests and goals who may be aware of grants or organizations who donate funds to these types of issues.

Research and grant funding is available for programs and continued maintenance for the garden. Consider applying for the grants with an existing nonprofit in the area in order to ensure community ownership of the project. Grants are typically geared toward issues such as nutrition, obesity, science education, agriculture, and community health. Grant funding can be found online for both urban agriculture projects and school garden projects. There is a website called the Rebel Tomato, started in Detroit, that focuses on youth gardening that has a wealth of resources managed by the American Community Gardening Association with funding from the USDA community food projects grants. Many school grants are up to $ 1,000 for nonprofits but some are available from $5,000 and up to $60,000 via the online resources of the USDA.

More than likely, a combination of financing resource capacities and capital input options will be utilized for most urban agriculture developments.

EXAMPLES OF CITY AND GOVERNMENT GRANTS

1. New York City: Farm City Fund is a new loan fund for urban farms and related businesses that support urban agriculture in the NYC metropolitan area.

    Loan amounts will range from $1,000 to $30,000.

    Interest rates will be 10% or less.

    Terms will be less than one year.

    Fewer than five (5) loans will be made in Year 1.

2. Canada: Tree Canada: Planting Fruit & Nut Trees for the Community. The purpose of the Edible Trees program is to offer funding of up to $4,000 and other resources for community-based projects that provide residents with access to fresh fruit and nut trees while making a positive difference to the Canadian environment including:

    Provide shade

    Absorb and deflect solar radiation

    Improve air quality

    Absorb and filter water

    Create habitat for wildlife

    Funding is available for, but not limited to, community gardening groups, community housing projects, schools, parks, and arboretums.

3. USDA's National Institute of Food and Agriculture (NIFA) manages the People's Garden Grant Program (PGGP), with funding from the Agriculture Marketing Service, Animal and Plant Health Inspection Service, Food and Nutrition Service, US Forest Service, and the Natural Resources Conservation Service. The grants announced, totaling $725,000, are the first awards given under the PGGP. USDA received more than 360 proposals requesting more than $4 million.

    PGGP was designed to invest in urban and rural areas identified as food deserts or food insecure areas, particularly those with persistent poverty. In addition, PGGP seeks to address health issues closely related to malnutrition, including food insecurity, obesity, diabetes, and heart disease, through onsite education programs. Projects in 2011 were funded in Alaska, Arizona, California, Colorado, Connecticut, Hawaii, Maryland, Michigan, and Ohio.

The Urban Food Jungle: An Adaptive Solution Idea Prototype

The Urban Food Jungle is a conceptual design that responds to the threat of diminishing food security and the inefficiencies of classic urban farming models by interconnecting sustainable food production, entertainment, education, and culinary delight (Figure 6.11). It has been envisioned as a high-yield design that has the potential to be deployed throughout various cities—addressing food scarcity through "positive impact design"—not just reducing the impacts of what we do, but positively giving back to the ecosystems to which we belong. The "aquaponic" system not only grows organic fruit and vegetables, but freshwater fish too—one nourishing the other in a sustainable, high-yield system. A series of pools (Figure 6.12) are used to raise the fish, generating nutrient-rich water as a by-product. This is circulated to the top of dramatic sculptural columns, fertilizing a variety of plants as they filter and clean the water on its descent back to the ponds, creating a lush edible canopy.

Figure 6.11 The Urban Food Jungle design is meant to draw people to the Urban Food Jungle to experience food in a variety of ways that include entertainment.

Figure 6.12 The series of fish pools and sculptural food columns can be designed in a variety of scales and patterns depending on its location and activities.

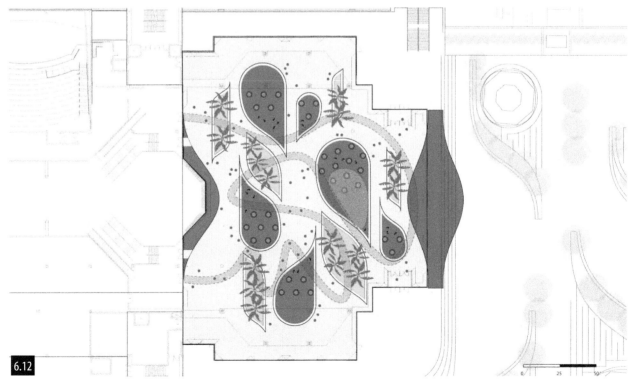

One pound of fish yields up to 55 pounds of produce, with the fish food also being a by-product of the plant production. This closed-loop system not only eliminates the need for artificial fertilizers, but uses up to 80-90 percent less water than traditional agriculture. Because many of our favorite fish are carnivores that consume more protein-rich food than they produce, they are inefficient to farm. Also, farmed-raised fish are often low in healthy omega-3 due to the amount of corn they are fed. In the Urban Food Jungle scenario, fast-growing omnivores such as Tilapia are raised and fed with excess vegetables, worms from wormeries used to digest organic waste from adjacent restaurants, and omega-3 rich flaxseeds.

In this particular application, the Urban Food Jungle inhabits a glazed winter garden adjacent to densely populated buildings, cafés, and restaurants. Ground-level pedestrian circulation enables easy visitor access; meanwhile, a floating pod-shaped food kiosk (Figure 6.13) serves fare prepared with fruit, vegetables, and fish cultivated on site—a focus for culinary demonstrations.

Most importantly, food harvested (Figure 6.14) from the Urban Food Jungle can be used to supply local restaurants, cafés, and farmers markets, feeding the immediate population. The system optimizes the use of direct natural light in two ways: first by strategically locating plants in areas with micro-climates that suit each species' specific needs; and second by supplementing light already being used by a building's occupants.

The Urban Food Jungle designers believe that this prototype has the potential to become a decentralized network of food production. It is an adaptable system, lending itself to be modified and embedded within the existing urban fabric in a variety of forms, including in double-height lobbies, sky gardens in tall buildings, and performative outer skins or green walls on buildings and other structures. They also believe it could be adapted to plazas, parks, and other outdoor spaces. As such, they believe it extends beyond low-impact design as it rehabilitates and responds to the existing urban fabric in a strategic and sustainable manner, which allows it to maintain the intrinsic efficiencies of urban density. They see the Urban Food Jungle as a model that will make it economically beneficial to use urban space to produce food. By directly linking the Urban Food Jungle to entertainment, restaurants, and education, its program becomes deeply embedded in mechanisms that will perpetuate its growth in a variety of social, spatial, and economic sectors. The Urban Food Jungle creates a dramatic and playful setting for children and adults alike to discover what the ingredients of their favorite foods look like, from plant to plate, informing food choices and reconnecting the urbanite with the productive landscape.

Figure 6.13 The winter garden uses edibles as a sculptural element to be used in the garden's markets, restaurants, and cafés. It is envisioned here as a culinary adventure.

Outreach and Community

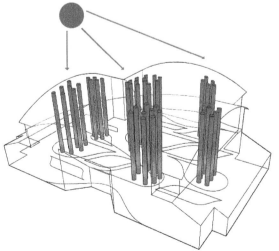

Figure 6.14 Using the greenhouse and hydroponics system allows for more crops than could be grown outdoors and also provides year-round food crops.

Design Team:	
Landscape Architect:	AECOM: James Haig Streeter, project director, Aidan Flattery, Blake Sanborn, Haein Lee, Jeremy Siew
Horticulture:	Eden Project UK

Forming Partnerships

Funding is an entrepreneurial venture, so expanding the network beyond the perceived boundary is another way to open up potential funding streams. Looking at how benefits can be maximized through partnerships such as a reduction in risk for both urban farmer and a restaurant, providing cheaper access to organic food for local consumers, or adding to the economic resiliency of the local community are all benefits that provide larger outcomes than just to the immediate users.

For example, what if there was an opportunity to partner with a city to measure the ecosystem services that the landscapes are providing in order to promote further investment in urban ag food landscapes throughout the city? Since it costs municipalities a lot of dollars to maintain their infrastructure systems such as maintaining clean air and water, managing their watersheds from runoff pollution, or finding solutions to deal with rising medical costs resulting from toxins in the environment, by documenting the metrics of these maintenance items through research grants or academic programs presents an opportunity for data synergy on both sides. Likewise if you can measure the benefits of an urban agricultural landscape that uses sustainable practices that aid in providing solutions to those maintenance issues of a citywide network, the possibility exists for the city to reallocate the money it is spending to more ecological land management techniques based on a sustainable landscape approach. As cities begin to respond to climate policy and greater environmental challenges, they will be searching for proven eco-management ways that they could adapt to. An academic partnership with a local university program would be one way to approach this type of research. We will discuss this further later in this chapter under the heading Monitoring and Research.

Other options include partnering in an economic relationship such as with restaurants that may be looking for local food branding opportunities that are urban agriculture based or develop a strong locavore tie in. This would result in a partnership beneficial to both parties such as an urban farm growing food exclusively for that restaurant or food retailer or other arrangements that fit with both companies' vision. This can build some sense of stability to the urban farm from an operational side, which generally is one of the chief drawbacks to a farm, and it builds stability for the retailer to have a source it can count on. Both companies would be able to benefit by the green job and cross branding aspects that are created by this partnership.

In residential neighborhoods, another option to explore is whether an urban ag landscape development be formatted into its own CSA or expanded to include a series of neighborhoods into a larger CSA. Investigations could also consider looking into the potential for an urban food landscape or farm to be located on available commercial property that is not being utilized in the community. This could be a lucrative exchange if the production math works out for both parties.

In various towns and cities, nonprofit organizations are being created to help foster urban agriculture within those areas to help citizens help themselves. One example of this type of organization is in Detroit, called GrowTown, founded by concerned landscape architects and planners who saw a need to fill in their underserved neighborhood communities in Detroit who have been seriously impacted by the latest recession. From its website: "GrowTown is a nonprofit organization dedicated to enabling neighborhoods, left fragmented in post-industrial cities and landscapes, to self-organize. Through grassroots community-driven design and local leadership, the Local Food Economy is the catalyst for growing resilient and sustainable neighborhoods that can respond to the important challenges of our time" (http://growtown.org/blog/).

Forming partnerships is also beneficial for education and outreach potential. There are a number of philanthropic organizations and nonprofits within communities that may prove beneficial as funding sources for capital funds, educational funds, and job training funds. Partnerships with organizations even include food banks or other charitable organizations that can benefit from excess produce, feeding more of the community, which also takes care of a percentage of urban agriculture waste management solutions. The partnerships do not necessarily just rely on funding but could supply management, leadership, and other needed resources. When forming partnerships, think about how the relationships can become mutually beneficial.

Expo 2015, Milan, Italy

Expo 2015 is a universal exposition to be held in Milan, Italy, with the theme of "Feeding the Planet, Energy for Life" (Figure 6.15). The expo's Conceptual Masterplan presented in July 2009 was designed by the Architecture Advisory Board consisting of Stefano Boeri Architetti, Richard Burdett, and Herzog & de Meuron, with sustainability guidelines developed by William McDonough + Partners. The further development of the project was realized by a planning team featuring 16 young and upcoming designers and engineers.

Figure 6.15 The universal expo "Feeding the Planet" is meant to demonstrate varied techniques and celebrations of urban agriculture around the globe.

240 Designing Urban Agriculture

Figure 6.16 The site, oriented along a central axis, will feature specialty areas highlighting particular topics and issues related to urban agriculture.

Figure 6.17 Different pavilions will showcase the techniques and flavors from around the world in a celebration of food and culture.

The team developed a compelling concept appropriate to the Feeding the Planet theme on the 271 acre site: 30-meter-wide strips of land are to be allocated to each of 130 participating countries, each fronting a central boulevard (Figure 6.16). The plots will incorporate demonstrations related to food growing, production, and preparation. Structures and pavilions are proposed to be demountable and constructed from safe materials that are designed to return to either biological or technical metabolisms. Participants are asked to interpret the expo's food-based theme in strong and original ways, and are encouraged to demonstrate their country's agricultural processes and technology, food production processes, and innovation at any stage of the food production chain (Figure 6.17).

Scheduled workshops and debates will cover topics such as improving food quality, security, and availability for all human beings, tackling and preventing health epidemics, innovation in the food supply chain, nutrition education, healthy lifestyles, and the value of cultural and ethnic heritage in culinary traditions. Additional exhibitions will include a restored historic local farm, and six hectares of agro-ecosystems in outdoor fields and greenhouses, which will allow visitors to "retrace the process by which human beings, compelled to search for food, learned to understand nature and thus how to transform it."

Architecture Advisory Board:

Herzog & De Meuron
Stefano Boeri Architetti
Richard Burdett & The London School of Economics
William McDonough + Partners

Finance 101

There are a few issues relating to determining whether to form a business enterprise. One of the first business decisions to determine is whether the urban agriculture enterprise is going to be a for-profit or nonprofit organization. Is the urban agriculture landscape large enough to support forming a commercial business, or should it be developed as a nonprofit organization business? Business enterprise management decisions require an advisory board or board of directors, a stakeholder committee, or another type of organizational leadership structure for the development of the operations management criteria and goals.

Decisions need to be made on approving annual budgets and approving funding sources that fit the project scope. Decisions also include oversight of the enterprise's finance/accounting records and budgets for marketing/outreach programs of the project.

PRODUCTION MANAGEMENT DECISIONS FOR EXCESS PRODUCE DISTRIBUTION INCLUDE:
- Local community donations
- Food banks
- Charitable organizations
- School lunch programs
- Personal use/employee benefits
- Barter systems
- Farmers markets and retail sales
- Monthly or weekly CSA deliveries
- Pay by affordability options
- Connecting with restaurants

Marketing and Educational Outreach

Marketing and educational outreach are important for community awareness, promoting environmental and human health benefits, promoting the integrity of the system, fundraising, green job creation, food safety, food security, and more. Branding is an element that creates identity in the marketplace (Figure 6.18). We discussed in the previous chapter that the marketing plan sets the framework for the communication network for outreach. Outreach communicates the project's value-based vision and benefits. These may include educational programs and job training or mentoring programs. Outreach to schools is typically an important focus and often includes special after school programs, summertime programs and year round seasonal events. The marketing outreach communicates how the vision strategies have become physical realities and the marketing message is aligned with the mission/vision of the urban food landscape. Channeling the ecoliteracy programs, and longevity stewardship programs into building the branding message is part of the marketing of urban ag landscapea.

In the new food economy, markets such as these food landscapes provide equal new risk for all involved, and need deep, trusting relationships and partnerships that you can count on for consistent sales and retail market strategies. These relationships will help the partners move through these changes together and encourage each other not to give up. If there is a culinary component, the chefs can help market the food message.

The first two questions when developing a marketing program for an urban ag landscape are:

1. What is the brand of the urban ag landscape that the community will identify with most and perceive that it also adds value to both individuals and the collective?
2. How will the marketing of this brand help establish credibility and avoid greenwash?

Figure 6.18 Branding and marketing are big ways to garner participants for a project and build identity.

The elements for marketing food landscapes are as follows:

- Developing the brand
- Marketing the brand
- Maintaining the brand
- Community outreach and education
- Websites
- Bulletin boards
- Blogging and social media
- Film and digital documentation
- Expanding the outreach
- Annual themed events for funding

Urban agriculture is part of today's green marketing that appeals to an audience with heightened environmental and social consciousness so traditional or conventional marketing methods are not an effective tool any more. The green marketing paradigm requires a more holistic and complex eco-conscious approach that recognizes the psychological and sociological shifts in today's green consumers—especially the connections between the environment and human health. Urban agriculture goes one step further by embodying the connections between food, culture, the environment, and human health.

An urban ag brand should incorporate the following strategies:

1. Convey the vision authentically
2. Appeal to audience's beliefs and values
3. Engage the local community
4. Educate and empower
5. Be credible
6. Communicate the ecological and sociological messages
7. Utilize third party partnerships that add to value

One of the first steps is to use the project's sustainable goals in a visual manner (Figure 6.19) as the framework for the marketing concept. Highlight the system strategies that were harnessed and will continue through the operations of the project.

These strategies might include any of the following:

1. Sustainable land use development practices
2. Resource protection
3. Recycled content and sustainable construction practices
4. Waste reduction: recycle, compost, zero waste
5. Organically grown
6. Reduction in toxicity
7. Energy efficiency
8. Water efficiency
9. Energy creation
10. Green job creation
11. Education and job training
12. Health and wellness

Figure 6.19 Providing information on your process and goals will allow people to better understand and be enthusiastic about the product. At Medlock Ames, photographer artist Douglas Gayeton depicted the winemaker Ames Morrison of Medlock Ames Winery in ways that express the winery's vision.

The marketing and educational outreach should consider how to partner with architectural, environmental education, science, and agricultural degree programs to gain support in maintaining the viability and visibility of the urban ag development. Also, this is another way to weave the local community into the project's sphere of influence in order to gain a sense of ownership and support from the community. Courses could be created with students and existing professors that tie into their current curriculum.

Another opportunity for marketing outreach is to look to the local government agencies to find champions that support local food economic development and green jobs programs. There are typically local congress people who might be hoping to write this type of legislation but are in need of success stories that would support it. By checking local news outlets and talking to city policy advocates you might be able to build an advocacy outreach network that could be harnessed for marketing and educational purposes. Through tapping into existing lobbying, nonprofits may try to connect their edible garden or farm to the lobbyist's own efforts to encourage the change they are trying to make while potentially helping the nonprofit urban become funded and more connected to the community.

Social media is another avenue and tool for promoting the message. It is easy, fast, and offers enormous potential in establishing a brand quickly and succinctly. It is also a great way to connect with the community you want to reach and for announcing events and programs. Websites such as Kickstarter have funded great ideas through proposals with a positive message that tap into the social zeitgeist. Kickstarter.com funds creative projects such as documentaries, but food landscapes could come under this category, perhaps in the form of creating a way to promote the ecoliteracy that is being accomplished by the food landscape. Marketing can also harness a social change message and use social media to help spread the story and brand in ways one article doesn't provide. The key is to build a communications network that is flexible, informative, and provides a perception of beneficial convenience to the audience.

Monitoring and Research: Tracking Value, Developing Incentives, and Urban Ag Metrics

Developing incentives for landscapes that benefit community and ecology requires research and monitoring. Urban agriculture landscapes offer an opportunity to partner with city entities and academic institutions to document and record the ecological, social, and economic benefits in progress. These metrics could begin to demonstrate what the potential savings are for municipalities and be used to inform the public to garner support for change. This research could address the costs to do job training, stormwater management, watershed management, economic development, community restoration, food security, waste management, climate adaptation, and even brownfield remediation.

To track value-added benefits, begin by measuring and tracking the ecosystem services such as stormwater, water conservation, CO_2, bioremediation, and more. Nature provides a wealth of ecosystem services to cities that far exceed the GDP. Urban agriculture is part of the solution for tying ecosystems into the green infrastructure of a city. Rating systems such as SITES, The Sustainable Sites Initiative, and LAF's Landscape Performance models are helpful to start looking at ecosystem metrics being collected. Urban agriculture should be included in sustainable sites metric research that links ecosystem health with community and city health.

> ### THE STORY OF HOW ONE BLOG POST BEGETS MORE MEDIA EXPOSURE
>
> 1. ASLA Sustainable Design and Development blog:
>
>   The Miller Creek Edible Garden appeared in this blog as an article: "The Power of One—It Takes Only One Student to Inspire a Village."
>
>   The story of a student, a garden, and the national obesity epidemic. An 11-year-old middle school student, Gabby Scharlach, inspires the fight against childhood obesity and promotes environmental stewardship with an organic garden classroom.
>
> 2. Teens Turning Green blog:
>
>   The article "Power of One" and a brief write-up of Gabby was reblogged in the Teens Turning Green blog, which was then contacted by Jamie Oliver's group Food Revolution.
>
> 3. Jamie Oliver's Food Revolution blog:
>
>   Jamie Oliver's group contacted Gabby to write her own article on the story of her garden for Food Revolution's weekly blog.
>
> 4. Lunch wars—Two Angry Moms book and blog:
>
>   Amy Kalafa contacted Teens Turning Green executive director Judy Shils after she read the Jamie Oliver piece by Gabby, and then re-blogged it on her blog, Two Angry Moms. Amy contacted Gabby and included "The Power of One" in her book *Lunch Wars* along with doing an interview with Gabby at a book signing in Berkeley for the book. Amy told Gabby that her story had inspired a 12-year-old girl on the East Coast to start a garden at her school.
>
> 5. Diggin Food blog:
>
>   This blog by a foodie in Seattle wrote a story about Gabby and her chef event that was held in the Miller Creek Garden and linked back to Food Revolution and the Miller Creek blog.
>
> 6. Miller Creek Edible Garden blog:
>
>   Gabby's own Miller Creek Edible Garden blog that she created and managed for her middle school also reposted links to all of these postings, creating a cross communications platform. This in turn gained her school and edible garden more notice by local politicians and the surrounding community.
>
> So, every bit of media coverage can get the story out and serve as inspiration to others. With social media the stories also gain traction and more exposure through the other web sites or other media it is linked to or re-blogged to.

Research could be linked to community issues such as urban pollution that food landscapes can remediate Do you take for granted that tap water is safe to drink? In too many communities, toxic chemicals from industrial waste and industrialized agricultural pollution transform kitchen faucets into health hazards. Rural and urban homes are left thirsty by contaminated community water, forcing California families and individuals to make a choice: buy bottled water, or drink toxic chemicals. The University of California at Davis recently published a report on the human health impacts of agricultural nitrates in drinking water, including blue baby syndrome and skin disease. Researchers found that *1 million Californians have been exposed to groundwater contaminated with nitrates in the past decade*. Forcing whole communities to buy bottled water to avoid serious illnesses is not an acceptable alternative to providing safe, affordable tap water.

Urban agriculture that focuses on community resiliency though education and job training will impact the local community with green jobs and new businesses that would grow out of the programs. These impacts should be tracked by cities as economic data that helps to foster the continuation for programs and business models that support the growth of resilient local economies.

There are currently a number of programs doing research on urban agriculture, sustainable farming practices and other aspects that include urban agriculture. Many of these are associated with universities or being run by nonprofits with grant subsidies. A few of these research programs have been highlighted as examples of the research network potential to tap into.

- The Urban Agriculture Research and Practice in Greater Philadelphia web site contains links to reports, articles, and policy papers related to urban agriculture in the greater Philadelphia area, including a link to University of Pennsylvania student work by graduates and undergraduates. Since 2007, professor Domenic Vitello and lecturer Michael Nairn have worked with students and colleagues to document production and distribution of food from community gardens and urban farms to examine the links between urban agriculture and community food security; prepare graduate and under graduate students to work in urban agriculture and food system planning; support city governments in developing equitable food and agriculture policy; and support urban agriculture organizations in developing their programs and capacity (Urban 2012).
- At Yale University, *The Yale Sustainable Food Project* created in 2003 manages the organic farm on campus and runs diverse educational programs that support exploration and academic inquiry related to food and agriculture. Yale University offers a degree in Environmental Studies with a Concentration in Sustainable Agriculture (Yale 2012).
- "Nature in the City" is San Francisco's first and only organization wholly dedicated to ecological conservation, restoration, and stewardship of the Franciscan bioregion. Directed by Peter Brastow their strategies and program areas are public education, community organizing and stewardship, ecological restoration, conservation advocacy and policy, and collaboration (Nature 2012).
- The *Earth Island Journal* is a quarterly magazine that combines investigative journalism and thought-provoking essays that make connections between the environment and contemporary issues, such as urban agriculture and its effects on health and food safety (Earth 2012).
- The National Sustainable Agriculture Coalition (NSAC) is an alliance of grassroots organizations that advocates for federal policy reform to advance the sustainability of agriculture, food systems, natural resources, and rural communities. NSAC's vision of agriculture is one where a safe, nutritious, ample, and affordable food supply is produced by a legion of family farmers who make a decent living pursuing their trade, while protecting the environment, and contributing to the strength and stability of their communities (National 2012).

- Full Circle Farm located in Sunnyvale, CA, was created as an Urban Agriculture Research and Demonstration project. As a part of the Human Tractor Society, this support network of people is committed to intensive food production in urban environments. Their work introduces the volunteer to biologically and human-intensive agriculture, including the Modified Raised Bed System (MRBS) that is featured in their quarter-acre research farm. The project is under the leadership of Wolfram Alderson, Executive Director of Full Circle Farm, and Jacob Morton, Urban Agriculture Specialist (Full 2012).

To address issues of food safety and food security in our cities there is a real need to document what is working and what is not working in city food sheds. In order to become part of the interconnected systems within a food system we need to begin by documenting and evaluating what exists, where the connections are and where new connections need to be designed for. Most cities do not see that the food sheds that create the food system are a part of their city infrastructure just as the water sheds of the water system currently are managed and documented. Documenting the metrics can add to the understanding of how the food system flows can better operate within the city. City Slicker Farms created a sustainability graphic in their annual report to the stakeholders that documents the metrics in a value based, community-oriented evaluation to communicate their "whole systems" approach. In this manner they used the metrics as a tool to market the message in a meaningful way (Figure 2.20).

More rigor is required to analyze and determine the impacts of urban agriculture on our culture, our cities, and our economy. Through white papers and case study observations the story will unfold as urban agriculture evolves and impacts our local communities. Partnering with research and academic institutions is one method in documenting the research and metrics relating to urban agriculture.

One example of this is symposiums and community charettes on urban agriculture that can focus a more intense lens on the issues. In 2010, Seattle declared it was the Year of Urban Agriculture as a way to explore and expand on its vital culture of community gardening, farmer's markets, and regional farming. At the University of Washington's College of the Built Environment, landscape architecture students organized a community charette for a two day event (Figure 6.21) to explore the role of design in the urban food movement.. Events such as these begin to expand the dialogue on how urban agriculture impacts our culture and our cities.

The event began with a studio presentation and a panel discussion to launch the two-day charette. The goal of the charrette was to tie into the mayor's challenge for 100 edible landscapes to be installed throughout the city that year. The panel included UW landscape architecture professor of history, Dr. Thaisa Way; Jason King, principal of TERRA.fluxus in Portland and author of popular blog Landscape + Urbanism; Jeff Hou, the chair of the UW Landscape Architecture department; Keith McPeters, a principal landscape architect at Gustafson Guthrie Nichols in Seattle; and Deb Guenther, a principal landscape architect at Mithun.

According to the charette participants, Jason King's presentation, which kicked off the charette, revealed Portland's long-time commitment to urban farming along with a series of five principles that he felt were necessary for cultivating urban agriculture:

1. Utilize a hierarchy of urban spaces.
2. Work through policy barriers.
3. Reframe permaculture in a new lens.
4. Maximize efficiency per square foot.
5. Develop orderly frames.

All of the panel participants shared their firm's efforts and thoughts on creating urban agriculture in the city to set up the dialogue that followed. Deb Guenther revealed that folks at Mithun were personally involved in a multi-partner program that supports a new generation of farmers and a program called LettuceLink, using vacant lots for farming that serves a neighborhood scale. She pointed out that our role as city designers takes us beyond the design of food production environments to include issues of delivery, consumption, and waste.

Jeff Hou posed an inspiring challenge in the form of a question to the group, "Can an urban food revolution be designed?" and/or "Can urban food revolutionize design?"

Keith McPeters summed up the panel's discussion by quoting JB Jackson's criteria for whether design is good—is it ecologically wholesome, socially just, and spiritually rewarding? These types of symposiums are great at asking the tough questions, talking about the road blocks, learning about projects and programs you might not be aware of, and offering up idea seeds that may inspire the participants to potentially develop further in their work.

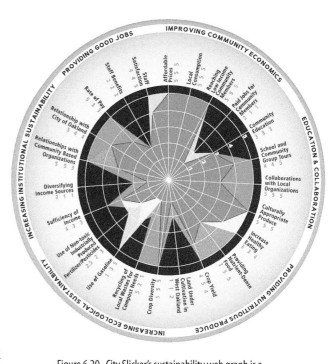

Figure 6.20 City Slicker's sustainability web graph is a values-based, community-oriented evaluation tool based on a "whole systems" approach.

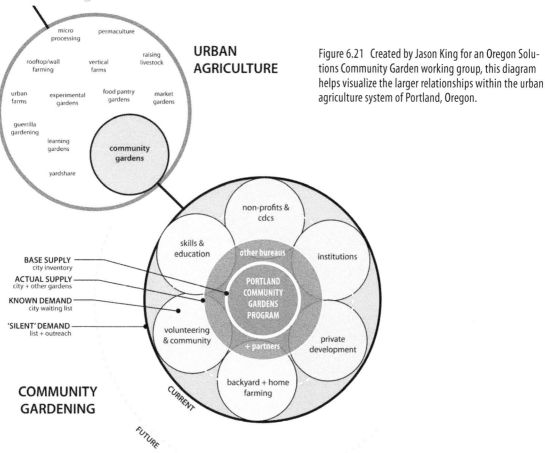

Figure 6.21 Created by Jason King for an Oregon Solutions Community Garden working group, this diagram helps visualize the larger relationships within the urban agriculture system of Portland, Oregon.

Alemany Farms, San Francisco, California

First planted under its current name in 2005, Alemany Farm sits on land in the southeast of San Francisco, which has hosted a variety of farming projects over the years (Figure 6.22). The largest property in San Francisco's Community Gardens Program, it is located on land owned by the San Francisco Recreation and Park department, with a small portion on the adjacent lot of the low-income housing development, Alemany Dwellings.

A portion of the farm's land is dedicated to 35 community gardens in which the city's residents can maintain their own plots, with additional plots dedicated to the residents of Alemany Dwellings (Figure 6.23).

The rest of the farm is used as a large demonstration garden and is maintained by a staff of volunteers. All of the farm's projects are managed by Friends of Alemany Farm, a collective of volunteer managers and workers (Figure 6.24).

Between 8,000 to 10,000 pounds of organic fruits and vegetables are produced by the farm each year. The harvest is split between volunteers, residents of Alemany Dwellings, and the Saint Martin de Porres foodbank.

The collective aims to improve the food network of San Francisco not just by providing produce to the community, but also by educating and training the public on agricultural matters (Figure 6.25). Community workdays are held twice a month, and field trips are offered to diverse groups, from kindergarten classes to corporate groups. The farm also hosts an 11-month internship for adults, which starts in March and takes participants through all of the seasons.

Figure 6.22 An overview of Alemany Farms and the adjacent neighborhood.

Outreach and Community 251

Figure 6.23 The garden is divided into plots available to city residents.

Figure 6.24 The garden is managed by workers and volunteers.

Figure 6.25 Alemany Farms aims to be a community center of education and production.

Design Team:

Clients: Friends of Alemany Farms; San Francisco Recreation and Park Department

Design: Friends of Alemany Farms

Policy and Advocacy

Cities have begun to take a new interest in urban agriculture as a way to promote health, to support economic and community development, and to improve the urban environment. Eaters alone cannot transform the food system. It is current public policy, not consumer choice, that props up the industrial food system and constrains the growth of organic farming. In fact, we have already learned that in many cities and towns across the country you cannot even grow food outside your door, and planting an edible landscape of any kind breaks local zoning codes. This section will talk about what is going on in San Francisco, New York, Austin, New Orleans, Detroit, Seattle, and other cities as it relates to setting new policies that allow for urban agriculture:

- Understanding current policies and codes in your area
- Understanding what groups are actively following upcoming legislation
- Partnering with universities and other educational opportunities to create pilot projects
- Planning strategies that promote and support urban agriculture and food sheds
- Tools and methodologies for integrating urban ag into city planning
- Sustainable guidelines that include food landscapes
- City and county ordinances that support and promote urban agriculture and the food system

Zoning and Policy

Urban agriculture projects provide neighborhoods with permanent or temporary amenities that contribute to a positive community image. Most often, these projects are responses to local food deserts, consumer demand, transit-lacking populations, and economic inequality. In other neighborhoods, they are created through a cultural demand or through a grassroots community process to serve a local desire and need. Because of its diversity of typologies and food shed components, urban agriculture can be seen as a powerful tool for urban planners, especially in rethinking the sustainable city.

Even if not intended, municipal policies can affect urban agriculture uses both negatively and positively. Some cities are promoting urban ag land uses through land donations, funding, or protective zoning. However, zoning can have a negative impact if the zoning is restrictive in nature. Zoning is typically more regulatory and restrictive in nature, so it is not always the best solution. Policies also can become a hindrance to food landscapes or food shed systems. Since many cities have old ordinances on their books that restrict urban agriculture, the first thing that urban planners can do is to review them and redesign ordinances that promote urban agriculture and the food shed system.

Comprehensive planning is a good approach to promote urban agriculture as a land-use planning issue. There are currently a number of cities such as Seattle that have developed comprehensive plans to include food landscapes as part of the open space park system, which is how the P-Patch community gardens have been supported. Seattle's 2005 plan requires at least one community garden for every 2,500 households in a neighborhood. Even with this support, Seattle's community gardens are so popular they have long waiting lists.

Seattle P-Patch Program, Seattle, Washington

Figure 6.26 P-Patch Interbay is a garden that is moving to its third location—a testament to its persistence.

Seattle has recognized that the benefits of community gardens extend beyond the personal gardener, to those passersby appreciating the beauty, to the attendees of the social events hosted at the garden, and has incorporated urban agriculture as a valuable open space and not just a production landscape. The P-Patch program intends for their gardens to be permanent installations to the city and receive the same protection as other parks and services enjoy, including protective zoning classifications. Efforts have been made to relocate and/or develop the community gardens on public lands as an assurance to their security. Funding is available through city grants programs, a Neighborhood Matching Fund, and private solicitation. The program itself is a staffed city program that facilitates the creation and maintenance of the gardens and focuses on community, youth, and market gardens, and food policy. Partnered with the city is the P-Patch Trust, which serves as the fiscal agent, funds accountant, fundraiser, and insurer for the garden parcels. The P-Patch Trust also assists in the arrangements for food bank donations and general communications for the P-Patch program.

As the largest community garden in the P-Patch program (Figure 6.26), the Interbay garden is a testament to persistence, having been located on three sites over its four-decade history, and advocacy, having successfully persuaded the local government of its civic importance.

Interbay P-Patch Design Team:

P-Patch and Gardeners (Interbay 1 and 2), Joe Neiford (Interbay 3), CAST Architecture (Interbay 3 sheds and kiosk)

In its current state, it sits on a one-acre parks department site and has 132 plots (Figure 6.27). Located adjacent to a busy commercial street, it is separated from the traffic by a long, landscaped berm and orchard. The onsite structures were built and are maintained by volunteers (Figure 6.28) (Figure 6.29). Many of the materials used in the garden are recycled from other city projects or donated by private firms. The gardening community welcomes visitors to enjoy their garden and social events.

Figure 6.27 The current site is located on dedicated parks department land.

Figure 6.28 The site structures were designed by CAST Architecture.

Figure 6.29 The site structures anchor the social space of the garden.

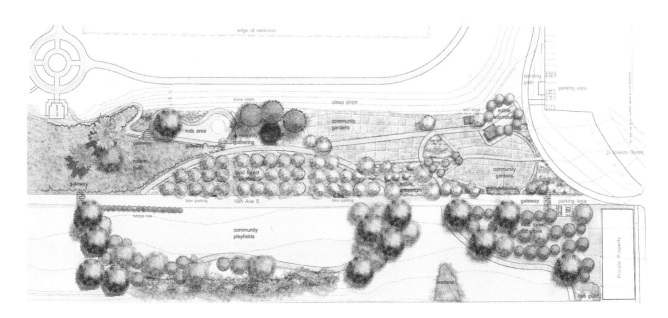

Figure 6.30 Beacon Food Forest approaches the urban agriculture installation as a fluid and ecologically sound parkland.

The Beacon Food Forest is an unconventional forest garden under development on a seven-acre site belonging to Seattle Public Utilities, in the Beacon Hill neighborhood (Figure 6.30). A grant was secured from the city of Seattle, and landscape architects Harrison Design were brought in to help lead a community design process (Figure 6.31) and develop a two-phase plan for the food forest's installation.

Using the principles of permaculture, the forest will employ a land management system that uses edibles to mimic a woodland ecosystem. The forest canopy will be composed of fruit and nut trees from around the world, an *edible arboretum,* while berry shrubs and other edible perennials form the understory. Companion plantings are incorporated to draw beneficial insects for pest management and soil amending, and to better the ecosystem's general health and productivity. Other site elements include a community garden in the P-Patch system and a children's area.

Beacon Food Forest P-Patch Design Team:	
Landscape Architect:	Harrison Design
City of Seattle:	Beacon Hill Community

Figure 6.31 The design process for the food forest made use of David Holmgren's Permaculture Principles.

Climate Protection and Sustainability Plans

Sustainability plans are another avenue to address urban agriculture and food issues. Some cities are making a connection to promote urban agriculture within their climate protection plans. The city of Baltimore's 2009 sustainability plan in its chapter on greening addresses the food shed, including food production and distribution in support of and provisions for urban agriculture opportunities (Baltimore 2009). And when Gavin Newsom, now lieutenant governor of California, was mayor of San Francisco, he issued a July 2009 executive directive that food system planning was the responsibility of city government. The city has followed up this directive with a number of policies that add to the promotion of food landscapes in all of its multi-dimensioned diversity in San Francisco.

Legislation and Policy Reform

Two recent policy changes in San Francisco included April 2011 legislation that created new zoning for urban agriculture and a July 2012 ordinance that sets goals and timelines for how the city government can better support urban farmers. The urban agriculture zoning legislation was spurred partly by a small urban farm called Little City Gardens who wanted to expand their plot. They petitioned for a re-writing of the code as they would have had to pay $3,000 in new permit use fees. Little City was the first to receive a zoning permit as a result of the legislation and the new permit only cost them $300.

The zoning legislation for urban agriculture included the following key points:

- Gardens and farms that are under one acre are welcome everywhere.
- People can sell the produce they grow in the city.
- New change of use permits were established for urban farms over one acre.
- It does not address animal husbandry.

The July 17, 2012, Ordinance goals and timelines go into effect:

- To complete and publish, by January 1, 2013, an audit of city-owned buildings with rooftops potentially suitable for both commercial and non-commercial urban agriculture;
- To develop, by January 1, 2013, incentives for property owners to allow temporary urban agriculture projects, particularly on vacant and blighted property awaiting development;
- To develop, by January 1, 2013, a streamlined application process for urban agriculture projects on public land, with clear evaluation guidelines that are consistent across agencies;
- To create, by July 1, 2013, a "one-stop shop" for urban agriculture that would provide information, programming and technical assistance to all San Francisco residents, businesses and organizations wishing to engage in urban agriculture;
- To develop new urban agriculture projects on public land where residents demonstrate desire for the projects, with at least 10 new locations for urban agriculture completed by July 1, 2014;
- To provide garden resource locations in neighborhoods across the city, at existing sites where possible, that provide residents with resources such as compost, seeds and tools, with at least 5 completed by January 1, 2014; and,
- To analyze and develop, by January 1, 2013, a strategy to reduce the wait list for San Francisco residents seeking access to a community garden plot to one year.

Eli Zigas, who is the Food Systems and Urban Agriculture program manager with

SPUR, helped to write the policy and the resource document put out by the San Francisco Urban Agriculture Alliance called "Starting a Garden or Urban Farm in San Francisco," which is available online. He also was responsible along with the Food Systems and Urban Agriculture Policy Board in putting out the latest SPUR report, Public Harvest, on how to expand the use of public land for urban agriculture.

In a conversation with Eli, he discussed a number of issues he felt were important with regard to setting policy for urban agriculture in the city. The first was that he believes that at the city level urban agriculture is not about the quantity of food. For most cities it would be impossible to feed its inhabitants only with food grown within the city. Even Havana whose story was told in Chapter 1 about its transformation to a true food economy can only grow 60 percent of the food it needs within its city limits. Detroit may be an exception since there is now so much vacant land available.

Eli discussed how SPUR spent six months developing the report Public Harvest, which includes a series of recommendations for how San Francisco can better capture the benefits of urban agriculture by expanding the number of projects on public land; outlining the need for better coordination among city agencies; more efficient use of existing funding; and greater access to public land. SPUR supports urban agriculture because of the multiple benefits to San Franciscans and the city as a whole. The report outlines the key benefits of urban agriculture as:

- Connecting the city residents to the broader food system
- Providing greenspace and recreation
- Saving public agencies money
- Providing ecological benefits and green infrastructure
- Building community
- Offering food access, public health, and economic development potential

What SPUR believes success looks like in San Francisco would mean from a resident's perspective that more San Franciscans will have access to space where they can grow food and from the city's perspective it would mean a more efficient and streamlined approach to providing support and public land to residents and community groups that want to support projects. Indicators of success that the city has gotten better at capturing the benefits of urban agriculture would include:

- Residents waiting no longer then a year for access to a community gardening plot or communally managed garden space.
- New urban agriculture projects launched on public land where residents demonstrate the desire for the projects.
- The creation of a "one-stop-shop" which would provide information, resources, and technical assistance for urban agriculture in the city including a single application for starting a new project that could be run by the city or a non-profit.
- More efficient use of public funds dedicated to urban agriculture, including lower costs for creating new sites and less duplication among city agencies.

For the entire report, see spur.org/publicharvest.

These types of policy changes will also need to include public policy reform. Overall in the U.S. there is a widespread lack of infrastructure for supporting small-scale ecologically minded farmers. The industrial food system siphons off public resources in the form of commodity subsidies and through the monoculture corn ethanol industry. Small-scale farmers have to pass on the costs of ecological stewardship to consumers who have been willing to pay with their forks, yet these farmers still cannot make a decent salary.

Meanwhile, industrial-scale farmers are allowed to generally trash the environment and be subsidized at the same time. Public policy reform must begin to address human health, food security, food deserts, ecological services, and green infrastructure in a more robust and transformative way in how we think about and plan for the food system in our cities. By shortening the food supply chain, the city will be able to reap a long list of benefits: increased food security; easy access to fresh, healthy foods and more green jobs especially in underserved areas; more green space and recreation; greener infrastructure that provides ecological benefits; less pollution and waste; and reinvigorated local economies.

When considering policy changes, it is helpful to consider:

- What the possible urban agriculture typologies are that would be appropriate for the city.
- What the most desirable widespread outcome would be, and what would have the least controversial impact.
- Where this use should be specifically encouraged.
- What should be allowed but would need to be controlled.
- Who the participants are and how positive relationships could be fostered.
- Whether the citywide food shed can be mapped and what it needs to make it better serve the neighborhoods and communities within the city.

Controversial Policies Abound

You can't have a lifestyle trend such as urban farming or food landscapes without finding some controversy. Many cities and towns have old bylaws or zoning codes that prohibit a person from actually eating any food they grow in their own yard! Although some cities such as San Francisco, New York, Baltimore, Seattle, and Detroit have begun to change laws and policy in support of urban agriculture—and as this trend continues to thrive because of food safety and security issues, the growing foodie locavore movement, and urban hipster cred—citizens in other cities and towns have been threatened with jail time or fines for planting a garden or organic farm on their own property. Yes, it seems that vegetable gardens are just not good for you or the community in these towns. Who knew planting a garden or creating an urban agriculture landscape could be the sign of a budding citizen activist? Some people made the news recently because of wanting to live a healthier lifestyle and grow their own food but now are being treated like criminals for doing so.

One example is Julie Bass in Oak Park, Michigan. After her front yard was torn up to replace a sewer line, she decided to plant a vegetable garden instead of the lawn that was originally there. But instead of being praised for her decision, she has been cited for a misdemeanor code violation because the city says that the edible garden is not a "suitable" front yard choice. City code says that all unpaved portions of a site shall be planted with grass or groundcover or shrubbery or other suitable live plant material. According to the planning director, *suitable* means common, and since there are no other vegetable gardens in the city's front yards, the Bass garden is unsuitable. But, a quick look at Webster's Dictionary will tell you otherwise. Julie was threatened with 93 days in jail for growing a garden in her front yard.

In Vancouver Island, Dirk Becker of Lantzville, British Columbia, turned his 2.5 acre property, which had been a gravel pit, into a thriving organic farm. The Beckers were cited under the "unsightly premises" bylaw for having small piles of manure on their property. The piles were not visible and just part of the sustainable farm's operations. According to the post on Grist.org, the letter came on the very day 8,000 compost bins were distributed to residents in their community. So gravel pit = okay but beautiful organic farm with real soil = not okay.

Glide Church—Graze the Roof, San Francisco, California

Figure 6.32 The rooftop garden at Glide Church is aimed at empowering people to make healthy food choices.

Figure 6.33 The program develops awareness of the food system.

Figure 6.34 Low-cost materials were used to build the rooftop "beds" on the existing roof.

Graze the Roof is an edible, rooftop, community garden on the top of the Glide Memorial Church near downtown San Francisco (Figure 6.32). The mission of Graze the Roof is to educate the community about how to grow food in the urban environment, to empower people to make healthy food choices through deepening their understanding of society's food system, and to demonstrate low-cost container gardening (Figure 6.33).

The garden makes use of simple, lightweight building materials, many of which are reused. A variety of produce is grown using earthboxes, hydroponic planters, and planters made from shipping palettes and milk crates (Figure 6.34). The group also raises bees, and uses worms to make compost. Graze the Roof integrates with the church's afterschool and summer programs for youth, as well as provides regular gardening and cooking classes for the wider community. The garden is maintained by a staff of community volunteers.

Food Security and Food Deserts

West Oakland is a food desert of 25,000 people, 30 liquor stores, and no grocery stores. It takes local residents 2 hours on average to travel by public transit to the closest options. This fits with the definition of a food desert, where people do not have access to healthy food due to a lack of grocery stores combined with low-income residents who cannot afford the time or money it takes to overcome access barriers. The USDA reports that 23.5 million Americans currently live in food deserts (Ver Ploeg 2012). More alarming still, this includes 6.5 million children. Food deserts illustrate the lack of equity that the current market forces of our industrial nonurban food systems create. Grocery stores have to make certain profits to consider opening a store, and the low-income food deserts don't typically fit the business model. The decentralization of food production into cities can start to break down the centralized food economy and provide more options and competition into the market.

West Oakland has many challenges to overcome for a vibrant urban food system: toxic soil, little public land for farming, industrial landscape, lack of biodiversity and habitat, no markets for selling, low rates of education, and very few homeowners who can use their land as they wish. Many of the ecosystem services of a healthy ecosystem are gone, with little vegetation or common spaces with good soil, water runoff management, clean air, beneficial insects, habitat for birds and native plants, space for perennial food crops like trees, and so on.

In 2001, a group of community members got together to address this concern by growing their own food in vacant lots, the first one being Willow Rosenthal's donated lot. Interested members of the neighborhood volunteered to grow the food and then sell it at Center Street Farm Stand and share the rest for donations. The effort has grown to include over 100 backyard gardens, seven Community Market Farms, weekly farmstands, a greenhouse, and urban farming education programs.

One thing that stands out in Oakland's effort to fix food deserts and other food justice issues is the development of the Oakland Food Policy Council. Through the community-driven efforts of City Slicker Farms, the city and the other 20 seats on the council have learned to listen and work with a diverse set of key stakeholders that represent the great diversity within the food system. This type of council can help inform efforts across the city through the sharing of diverse perspectives that are rarely in the same room talking through dynamic change efforts, and by tapping into the wisdom of this group. This is a valuable learning community that can provide advice and holistic systems thinking for projects across the city with a perspective that is hard to bring together as an individual project without many key partnerships.

Gotham Greens, Brooklyn, New York City, New York

Figure 6.35 Gotham Greens brings year-round urban agriculture to the middle of New York city.

Figure 6.36 The rooftop site demonstrates a new use for what is usually nonproductive areas while taking advantage of sun exposure.

Gotham Greens is a commercial, climate-controlled, hydroponic greenhouse farm located on the roof of a warehouse in Greenpoint, Brooklyn (Figure 6.35). While working on the Science Barge project, Gotham Greens co-founder Viraj Puri became familiar with hydroponic growing systems, and saw the commercial potential of such a system. Puri and co-founder Nick Haley started Gotham Greens in 2008. Jenn Nelkin, a fellow Science Barge alum with extensive experience farming in unconventional environments, was brought in as the company's greenhouse and hydroponics expert.

Part of the founders' goal was to create a farm that thrives in its unique environment of New York City. With a near absence of available open space in the city, this meant turning to spaces that are available—rooftops (Figure 6.36).

Figure 6.37 The greenhouse allows for year-round production.

Due to issues of zoning and structural stability, it took over a year to find the right building on which to lease roof space. A custom greenhouse was designed and installed, along with a large solar array to help power the outfit (Figure 6.37). Unique among rooftop farms, the greenhouse component allows the growers to control temperature and humidity, enabling Gotham Greens to remain productive throughout the year.

In order to make such an operation more productive, and thus more economically sustainable, a hydroponic system was chosen to maximize yield. The system allows the farmers to precisely control the delivery of minerals and micronutrients to their plants. Combined with the climate-controlled greenhouse, this means the produce is consistent throughout the year. Hydroponics also eliminates the use of soil, decreasing building loads, eliminating runoff and erosion, and creating a more sterile environment to eliminate disease and pests (Figure 6.38). Pests that do find their way into the greenhouse are managed by the introduction of beneficial insects, such as ladybugs to control aphids.

The first year of operations at Gotham Greens was a great success, producing even more than the goal of 100 tons of produce in the form of leafy greens. This goal is more than twice the yield that a soil-based farm would produce, which has increased the economic return for an

Figure 6.38 Hydroponics provide the nutrients for the plants in place of soil.

urban farm. The farm's crops are in high demand among the city's supermarkets and restaurants, and the produce is known for its consistent quality and taste. Furthermore, since the plants are grown in the city, they can be harvested on the day of delivery, ensuring freshness and ripeness, which is hard to find in food shipped in from outlying agricultural areas. Additionally, the farm has provided green jobs for a workforce of over 20 local residents.

Though Gotham Greens has already reached a state of profitability, its founders are looking toward the future with an eye to expand to additional greenhouses, and to grow new crops like tomatoes. They advise that commercial rooftop operations are not always easy and must be planned very carefully, but that the most important thing is to farm in a way that is appropriate to the environment and context.

Design Team:

Designer:	Gotham Greens
Architectural Consultant:	New York Design Architects

The Challenge Ahead

The majority of humans now live in urban environments. If there was a natural disaster or any kind of event that could shut down a city's food supply chain, supermarket shelves would be emptied within a few days in most large cities. Thanks to our dependence on the globalized food system, most cities and their populace would not be prepared for such emergencies. How would people feed themselves? Would they resort to foraging from the city sidewalks or city parks? How would a city feed itself with the current food system in place, which does not foster resiliency? How will a populace that is low in ecoliteracy become more self-sufficient? How do we create conditions that allow for each community to feed itself? We are now losing farmland at the rate of more than 1 acre every minute. If farms disappear, food will disappear. What can we do to change this trajectory and focus our direction on a new food system model that will meet all of these challenges?

There are many bright spots of hope that are shining through in each area of the urban ag lifecycle process spheres that this book has highlighted. At the grassroots level these new urban ag projects, policies, ideas, organizations, and forums have the potential to light a spark and spread out into our urban and peri-urban communities as a growing network of urban agricultural systems and create more impetus for a new food system model. This will not happen overnight and it also requires us to not just buy local food from farmers markets, or design local food and farm projects, or create new urban ag policies, but also to cultivate, support, and facilitate the lifecycle aspects of this growing movement into a real sustainable food web wherever we can.

In the food world, for example, there is a new set of culinary guidelines starting to emerge of voluntary standards such as architecture's LEED certification called SPE, which stands for *sanitas per escam* (Latin for "health through food") and stresses not only using local, seasonal ingredients but also combining them in ways that maximize their nutritional value. Chef Jeremy Bearman, a Michelin star chef at his restaurant Rouge Tomate in Manhattan, is working with a dietician to make sure his haute cuisine is also healthy. However, the restaurant's owner and SPE creator Emanuel Verstraeten kept the good-for-you message off the menu until recently since most diners equate healthy with bland. Verastraeten is ready to go public with this culinary certification, convinced by the interest in its principles by other chefs. In spring of 2012 he launched a certification program that invites U.S. chefs to submit recipes or entire menus for revision to meet the criteria he has developed with the help of several nutritionists. His goal is for the SPE logo to become a selling point at restaurants. A few chefs have already signed on. Most believe that while a logo alone won't make people want to eat healthier, it is a good step towards making restaurant food better (Park 2012).

In the farm world, organizations such as the American Farmland Trust (AFT) are working on a national level to protect the nation's farm and ranch land—keeping it healthy and improving the economic viability of agriculture. They are raising awareness on the issues of the relationship of well managed agriculture and the environment, the value and need for growing local and supporting local farms, farmland protection, and Federal Farm policy that will be beneficial. They are also working at the state levels, which are a very good overall resource on the political issues, events, and initiatives that are and will affect your community. Getting more involved is part of the designer as change agent approach. Become an active local and global citizen.

Another example to take note of is the events and symposiums centered on the food system challenges that are cropping up around the country. Recently I was invited to an Eat Your Sidewalk Showdown event in San Francisco. *Eating your sidewalk*—that either sounds like a lot of fun or a very scary proposition depending on how familiar you might be with the term "urban foraging." This event, co-sponsored by Spurse, a research and design collaborative, and AECOM, was held in the SOMA area of San Francisco. It brought together farmers, chefs, environmentalists, developers, designers, city staff, and various intrepid urban adventurers to participate in a foraging cook off and discussion on the urban agriculture issues raised over the course of the all-day event.

The daylong event was broken into three parts: Forage, Cook, and Discuss. Led by Iain Kerr of Spurse, the group of foragers was tasked with finding ingredients in the surrounding neighborhood to add to the fresh ingredients for the cook-off challenge. Spurse calls this the "MacGyver the World" mentality of taking nothing at face value. The experience was definitely enlightening and what at first seemed almost impossible that anything other than dandelions would be found turned into a unique change of perspective on how a person could actually live off the land even in an urban forest when you discover that nature is urban. The simple act of looking for food in the city makes you look at the city differently and feel like you are a natural part of the city's systems just like the food you find on your journey. The city is a complex ecosystem.

The rules of the cook-off part of the event were that the chefs could use only the foraged materials, the fresh farm ingredients supplied by the local farmers, and they could bring 3 of their own ingredients. A surprise mystery protein would also be supplied by a local hunter that had to be used. The chefs were SF based Chef Takumi Abe and the Spurse team. The food that resulted from this challenge was deemed "earthy" and "tasty" and "I'd order this in a restaurant" worthy by the three judges.

The discussion afterwards ranged from the plight of food deserts and food justice to food security and food waste, the "Tragedy of the Commons" argument of whether a community can self-regulate or not, and much more. The group agreed that there was a critical need for a paradigm shift in society on how we grow, consume, and manage the food system in the city. Asked what do you think is the most important thing that needs to be done, most people said, "more education on the connection between food and health" and "making healthy food more convenient to all." I believe that this will require our relationship with food to change to one that sees food as an integral part of the web of the city fabric that is connected on a daily level to our own lives.

The need for an ecological food model has never been more needed in our cities than now. Let's decide to invite food back into our cities and forge a path towards creating healthier communities and a healthier environment by cultivating this intersection of ecology, design, and community through designing urban agriculture and a systems thinking approach. Where will we go, what will we do, and how will our cities change? Will the new farmlands be the rooftops of our buildings, the sidewalks of our streets, or any other combination of typologies in this book? How will this intersection help shape a more self-sustaining city in our future?

Resources

Baltimore Office of Sustainability. *Annual Sustainability Report.* 2009. www.baltimoresustainability.org/resources/index.aspx.

Earth Island Institute. "Earth Island Journal." 2012. www.earthisland.org/journal/.

Full Circle Farm. "Urban Agriculture Research & Demonstration Project." 2012. www.fullcirclesunnyvale.org/urbanagresearch/.

National Sustainable Agriculture Coalition. *Farmers' Guide to Value-Added Producer Grant Funding.* August 2012. http://sustainableagriculture.net/publications/.

Nature in the City. 2012. http://natureinthecity.org/.

Park, Alice. Food-Certifiably Good, *Time* magazine, July 2012.

San Francisco Planning + Urban Research Association. *Public Harvest: Expanding the Use of Public Land for Urban Agriculture in San Francisco.* April 2012. www.spur.org/publications/library/report/public-harvest.

Urban Agriculture Research and Practice in Greater Philadelphia. 2012. https://sites.google.com/site/urbanagriculturephiladelphia/.

Ver Ploeg, Michele, Vince Breneman, Paula Dutko, Ryan Williams, Samantha Snyder, Chris Dicken, and Phil Kaufman. Access to Affordable and Nutritious Food: Updated Estimates of Distance to Supermarkets Using 2010 Data, ERR-143, U.S. Department of Agriculture, Economic Research Service, November 2012.

Yale University. "Yale Sustainable Food Project." 2012. www.yale.edu/sustainablefood/.

Bibliography

Books

Abi-Nader, Jeanette, David Buckley, Kendall Dunnigan, and Kristen Markley. *Growing Communities Curriculum: Community Building and Organizational Development Through Community Gardening.* Philadelphia: American Community Gardening Association.

Allen, Will. *The Good Food Revolution: Growing Healthy Food, People, and Communities.* New York: Penguin Group, 2012.

Bateson, Gregory. *Steps to an Ecology of Mind.* Chicago: University of Chicago Press, 2000.

Berry, Wendell. *The Unsettling of America: Culture & Agriculture.* San Francisco: Sierra Club Books, 1996.

Bucklin-Sporer, Arden, and Rachel Kathleen Pringle. *How to Grow a School Garden: A Complete Guide for Parents and Teachers.* Portland: Timber Press, Inc., 2010.

Calkins, Meg. *The Sustainable Sites Handbook.* Hoboken: John Wiley & Sons, Inc., 2012.

Carpenter, Novella, and Willow Rosenthal. *The Essential Urban Farmer.* New York: Penguin Group, 2011.

Center for Ecoliteracy. *Big Ideas: Linking Food, Culture, Health, and the Environment.* Berkeley: Learning in the Real World, 2008.

Covey, Stephen M. R. *The Speed of Trust: The One Thing That Changes Everything.* New York: Free Press, 2006.

Creasy, Rosalind. *Edible Landscaping: Now You Can Have Your Gorgeous Garden and Eat It Too!* San Francisco, Sierra Club Books, 2010.

Danks, Sharon Gamson. *Asphalt to Ecosystems: Design Ideas for Schoolyard Transformation.* Oakland: New Village Press, 2010.

de la Salle, Janine, and Mark Holland, eds. *Agricultural Urbanism: Handbook for Building Sustainable Food Systems in 21st Century Cities.* Winnipeg: Green Frigate Books, 2010.

Ellin, Nan. *Integral Urbanism.* New York: Routledge, 2006.

Fox, Thomas. *Urban Farming: Sustainable City Living in Your Backyard, in Your community, and in the World.* Irvine: BowTie Press, 2011.

Frumkin, Howard, Lawrence Frank, and Richard Jackson. *Urban Sprawl & Public Health: Designing, Planning, and Building for Healthy Communities.* Washington D.C.: Island press, 2004.

Holman, Peggy, Tom Devane, and Steven Cady. *The Change Handbook: The Definitive Resource on Today's Best Methodsa for Engaging Whole Systems.* San Francisco: Berrett-Koehler Publishers, Inc., 2007.

Holmgren, David. Permaculture: *Principles & Pathways Beyond Sustainability.* Hepburn: Melliodora Publishing, 2002.

Haeg, Fritz, Ed. *Edible Estates: Attack on the Front Lawn.* New York: Metropolis Books, 2010.

Hou, Jeffrey, Julie M. Johnson, and Laura J. Lawson. *Greening Cities Growing Communities: Learning from Seattle's Urban Community Gardens.* Seattle: University of Washington Press, 2009.

Hou, Jeffrey. *Insurgent Public Space: Geurrilla Urbanism and the Remaking of Contemporary Cities.* New York: Routledge, 2010.

Hough, Michael. *City Form and Natural Process: Towards a New Urban Vernacular.* New York: Routledge, 1989.

Kalafa, Amy. *Lunch Wars: How to Start a School Food Revolution and Win the Battle for Our Children's Health.* New York: Penguin Group, 2011.

Kellogg, Scott, and Stacy Pettigrew. *Toolbox for Sustainable City Living.* Brooklyn: South End Press, 2008.

Kelly, Kevin. *Out of Control: The New Biology of Machines, Social Systems, and the Economic World.* Jackson: Perseus Books Group, 1994.

Kotter, John P. *Leading Change.* Cambridge: Harvard Business Review Press, 1996.

Kotter, John P. and Dan S. Cohen. *The Heart of Change: Real-Life Stories of How People Change Their Organizations.* Cambridge: Harvard Business Review Press, 2002.

Levesque, Matthew. *The Revolutionary Yardscape: Ideas for Repurposing Local Materials.* Portland: Timber Press, 2010.

Linn, Karl. *From Rubble to Restoration: Sustainable Habitats Through Urban Agriculture.* San Francisco: Urban Habitat Program of Earth Island Institute, 1991.

Louv, Richard. *Last Child in the Woods: Saving Our Children from Nature-Deficit Disorder.* Chapel Hill: Algonquin Books, 2008.

Madigan, Carleen, ed. *The Backyard Homestead.* North Adams: Storey Publishing, 2009.

Marcus, Clare Cooper. *Healing Gardens: Therapeutic Benefits and Design Recommendations.* New York: John Wiley & Sons, 1999.

Markham, Brett L. *Mini Farming: Self-Sufficiency on ¼ Acre.* New York: Skyhorse Publishing, 2010.

Meadows, Donella H. *Thinking in Systems.* White River Junction: Chelsea Green Publishing Company, 2008.

Mollison, Bill. *Permaculture: A Designers' Manual.* Tyalgum: Tagari Publications, 1988.

Montgomery, David. *Dirt: The Erosion of Civilizations.* Berkeley: University of California Press, 2012.

Orr, Stephen. *Tomorrow's Garden: Design and Inspiration for a New Age of Sustainable Gardening.* New York: Rodale Inc., 2011.

Ottman, Jacquelyn A. *The New Rules of Green Marketing: Strategies, Tools, and Inspiration for Sustainable Branding.* San Francisco: Berrett-Koehler Publishers, Inc., 2011.

Pollan, Michael. *In Defense of Food: An Eater's Manifesto.* New York: Penguin Group, 2009

Rich, Sarah. *Urban Farms.* New York: Abrams, 2012.

Solomon, Reggie, and Michael Nolan. *I Garden: Urban Style.* Cincinnati: Betterway Home Books, 2010.

Spiegelman, Annie. *Talking Dirt: The Dirt Diva's Down-to-Earth Guide to Organic Gardening.* New York: Penguin Group, 2010.

Standage, Tom. *An Edible History of Humanity.* New York: Walker Publishing Company, Inc., 2009.

Stone, Michael K. *Smart by Nature: Schooling for Sustainability.* Healdsburg: Watershed Media, 2009.

Steingraber, Sandra. *Living Downstream: An Ecologist's Personal Investigation of Cancer and the Environment.* Philadelphia: Da Capo Press, 2010.

Tracey, David. *Urban Agriculture: Ideas and Designs for the New Food Revolution.* Gabriola Island: New Society Publishers, 2011.

Waters, Alice. *Edible Schoolyard: A Universal Idea.* San Francisco: Chronicle Books, 2008.

Films

Connected. DVD. Directed by Tiffany Shlain. San Francisco: The Moxie Institute, 2011.

Dirt! The Movie. DVD. Directed by Bill Beneson. New York: Docurama Films, 2010.

Flow. Video. Directed by Irena Salina. New York: Oscilloscope Pictures, 2008.

Food, Inc. DVD. Directed by Robert Kenner. New York: Magnolia Pictures, 2009.

Hope in a Changing Climate. Video. John Liu. Xinxiang: Environmental Education Media Project, 2009.

Living Downstream. DVD. Directed by Chanda Chevannes. Toronto: The People's Picture Company.
Nourish: Short Films. DVD. San Francisco: Worldlink, 2009.
The Power of Community: How Cuba Survived Peak Oil. DVD. Directed by Faith Morgan. Yellow Springs: The Community Solution, 2006.

Blogs

American Society of Landscape Architects. *Sustainable Design and Development Blog.* http://sustainableppn.asla.org/
Carpenter, Novella. *Ghost Town Farm* (blog). http://ghosttownfarm.wordpress.com/
Dana Treat. http://danatreat.com/
Galloway, Willi. *Diggin Food* (blog). http://www.digginfood.com/about/
Garvin, Jennifer. *Starting With Dirt.* http://startingwithdirt.com/about/
Hellberg, Helge. *An Organic Conversation* (blog). http://www.helgehellberg.com/blog/
Holmes, Heather, Atlanta Botanical Garden. *Plant to Plate.* http://abgplanttoplate.blogspot.com/
King, Jason. *Landscape Urbanism* (blog). http://landscapeandurbanism.blogspot.com/
Solomon, Reggie. *UrbanGardenCasual* (blog). http://urbangardencasual.com/
Solomon, Reggie. *TomatoCasual* (blog). http://tomatocasual.com/

Websites

American Community Gardening Association. http://www.communitygarden.org/
Center for Agroecology & sustainable Food Systems, The. http://casfs.ucsc.edu/
City Farmer News. http://www.cityfarmer.info/
Community Food Security Coalition. http://foodsecurity.org/
Ecology Center of Berkeley. http://ecologycenter.org/
Edible Schoolyard, The. http://edibleschoolyard.org/
Education Outside. http://www.educationoutside.org/
Food Project, The. http://thefoodproject.org/
Garden Mosaics. http://communitygardennews.org/gardenmosaics/
Greenhealth Academy. http://practicegreenhealth.org/greenhealth-academy
Greenroofs.com.
Grist: Environmental News, Commentary, Advice. http://grist.org/
Growing Power, Inc. http://www.growingpower.org/
Inhabitat. http://inhabitat.com/
Institute for Food & Development Policy. http://www.foodfirst.org/
Jamie Oliver's Food Revolution. http://www.jamieoliver.com/us/foundation/jamies-food-revolution/home
Kitchen Gardeners International. http://kgi.org/
Land Institute, The. http://www.landinstitute.org/
Mother Earth News. http://www.motherearthnews.com/
National Farm to School Network. http://www.farmtoschool.org/
National Policy & Legal Analysis Network to Prevent Childhood Obesity, The (NPLAN). http://changelabsolutions.org/childhood-obesity
Natural News. http://naturalnews.com/
New York 2 New Orleans Coalition. http://www.ny2no.org/
Permaculture Principles. http://permacultureprinciples.com/
Resource Centres on Urban Agriculture & Food Security. http://ruaf.org/
San Francisco Urban Agriculture Alliance. http://www.sfuaa.org/
Seattle P-Patch Community Gardens. http://www.seattle.gov/neighborhoods/ppatch/
Slow Food USA. http://www.slowfoodusa.org/
Statewide Integrated Pest Management Program (University of California). http://www.ipm.ucdavis.edu/

Sustainable Agriculture Education (SAGE). http://www.sagecenter.org/
Sustainable Communities Online. http://sustainable.org/
Teens Turning Green. http://www.teensturninggreen.org/
Treehugger. http://www.treehugger.com/
Union of Concerned Scientists. http://www.ucsusa.org/
USDA List of State Soils. http://soils.usda.gov/gallery/state_soils/
Worldwide Permaculture Network. http://www.permacultureglobal.com/

Image Credits

Chapter 1
Figure 1.1 © Beth Hagenbuch
Figure 1.2 © Beth Hagenbuch
Figure 1.3 © Beth Hagenbuch
Figure 1.4 © Beth Hagenbuch
Figure 1.5 © Beth Hagenbuch
Figure 1.6 © Beth Hagenbuch
Figure 1.7 © Beth Hagenbuch
Figure 1.8 © Beth Hagenbuch
Figure 1.9 © April Philips Design Works
Figure 1.11 © Aidlin Darling Design
Figure 1.12 © April Philips Design Works
Figure 1.13 © April Philips Design Works
Figure 1.14 © April Philips Design Works
Figure 1.15 © City Slicker Farms
Figure 1.16 © City Slicker Farms
Figure 1.17 Courtesy of CMG Landscape Architecture
Figure 1.18 Courtesy of CMG Landscape Architecture
Figure 1.19 Photo by Malia Everette
Figure 1.20 © 2011 by Jeff Greenwald / EthicalTraveler.org
Figure 1.21 Photo by Malia Everette
Figure 1.22 © 2011 by Jeff Greenwald / EthicalTraveler.org
Figure 1.23 Photo by Malia Everette
Figure 1.24 © April Philips Design Works
Figure 1.25 Courtesy of Spackman Mossop Michaels
Figure 1.26 Courtesy of Spackman Mossop Michaels
Figure 1.27 Courtesy of Spackman Mossop Michaels
Figure 1.28 Courtesy of Spackman Mossop Michaels
Figure 1.29 © April Philips Design Works
Figure 1.30 Courtesy of Big City Farms
Figure 1.31 Courtesy of Big City Farms

Chapter 2
Figure 2.1 Courtesy of Prairie Crossing, photo by Michael Sands
Figure 2.2 Courtesy of Prairie Crossing, photo by Michael Sands
Figure 2.3 Courtesy of Sandhill Organics, photo by Peg Sheaffer
Figure 2.4 Courtesy of Hoerr Schaudt Landscape Architects, rendering by Bruce Bondy
Figure 2.5 Courtesy of The Farm Business Development Center, photo by Michael Sands
Figure 2.6 © April Philips Design Works
Figure 2.7 © April Philips Design Works
Figure 2.8 © April Philips Design Works
Figure 2.9 Top image: Frisbie Architects
Bottom image: With permission by Garrett Gill, Kari Haug, Gill Design, Inc., Landscape Architects
Figure 2.10 Copyright St. Croix Valley Habitat for Humanity 2011
Figure 2.11 © April Philips Design Works
Figure 2.12 © April Philips Design Works
Figure 2.13 © April Philips Design Works
Figure 2.14 © April Philips Design Works
Figure 2.15 © April Philips Design Works
Figure 2.16 © Hoerr Schaudt Landscape Architects
Figure 2.17 © Kenneth Weikal Landscape Architecture
Figure 2.18 © April Philips Design Works
Figure 2.19 © April Philips Design Works
Figure 2.20 © April Philips Design Works
Figure 2.21 Courtesy of William T. Eubanks, FASLA
Figure 2.22 Courtesy of William T. Eubanks, FASLA
Figure 2.23 SWA Group
Figure 2.24 SWA Group
Figure 2.25 SWA Group
Figure 2.26 SWA Group
Figure 2.27 by Joe Runco, SWA Group
Figure 2.28 SWA Group
Figure 2.29 © April Philips Design Works

Chapter 3
Figure 3.1 © Ries van Wendel de Joode
Figure 3.2 © Ries van Wendel de Joode
Figure 3.3 © Ries van Wendel de Joode
Figure 3.4 © Ries van Wendel de Joode
Figure 3.5 © Ries van Wendel de Joode
Figure 3.6 © Ries van Wendel de Joode
Figure 3.7 © Ries van Wendel de Joode
Figure 3.8 © Ries van Wendel de Joode
Figure 3.9 © April Philips Design Works
Figure 3.10 © April Philips Design Works
Figure 3.11 Courtesy of Latino Graduation Café in Arlington, VA, photo by Israel Salas
Figure 3.12 © April Philips Design Works
Figure 3.13 © April Philips Design Works
Figure 3.14 © April Philips Design Works
Figure 3.15 © April Philips Design Works
Figure 3.16 © April Philips Design Works
Figure 3.17 © April Philips Design Works
Figure 3.18 © April Philips Design Works
Figure 3.19 © April Philips Design Works
Figure 3.20 © April Philips Design Works
Figure 3.21 © April Philips Design Works
Figure 3.22 © April Philips Design Works
Figure 3.23 © Scott Shigley
Figure 3.24 © Scott Shigley
Figure 3.25 © Hoerr Schaudt Landscape Architects
Figure 3.26 © Scott Shigley
Figure 3.27 © Scott Shigley
Figure 3.28 © April Philips Design Works
Figure 3.29 © April Philips Design Works
Figure 3.30 © April Philips Design Works
Figure 3.31 © April Philips Design Works
Figure 3.32 SB Architects
Figure 3.33 SWA Group
Figure 3.34 © April Philips Design Works

Chapter 4
Figure 4.1 © Marion Brenner
Figure 4.2 © Marion Brenner
Figure 4.3 © Marion Brenner
Figure 4.4 Courtesy of Nelson Byrd Woltz

Image Credits

Figure 4.5	© Marion Brenner
Figure 4.6	© Marion Brenner
Figure 4.7	© Marion Brenner
Figure 4.8	© Marion Brenner
Figure 4.9	© Marion Brenner
Figure 4.10	© April Philips Design Works
Figure 4.11	© April Philips Design Works
Figure 4.12	Courtesy of Our School at Blair Grocery
Figure 4.13	Courtesy of Our School at Blair Grocery
Figure 4.14	Courtesy of Our School at Blair Grocery
Figure 4.15	Courtesy of Our School at Blair Grocery
Figure 4.16	Copyright St. Croix Valley Habitat for Humanity 2011
Figure 4.17	© April Philips Design Works
Figure 4.18	© April Philips Design Works
Figure 4.19	Courtesy of Rios Clementi Hale Studios
Figure 4.20	Courtesy of Rios Clementi Hale Studios
Figure 4.21	Courtesy of Rios Clementi Hale Studios
Figure 4.22	Courtesy of Rios Clementi Hale Studios
Figure 4.23	© Hoerr Schaudt Landscape Architects
Figure 4.24	© April Philips Design Works
Figure 4.25	© April Philips Design Works
Figure 4.26	© April Philips Design Works
Figure 4.27	Courtesy of Groundwork Hudson Valley
Figure 4.28	Courtesy of Groundwork Hudson Valley
Figure 4.29	© BASE Landscape Architects
Figure 4.30	© April Philips Design Works
Figure 4.31	© April Philips Design Works
Figure 4.32	© April Philips Design Works
Figure 4.33	Courtesy of Vitus Group, Inc., photo by Alejandro Garcia

Chapter 5

Figure 5.1	Photo by Judith Stilgenbauer
Figure 5.2	Bavaria Luftbilds Verlag GmbH
Figure 5.3	Rainer Schmidt Landscape Architects
Figure 5.4	Photo by Judith Stilgenbauer
Figure 5.5	Photo by Judith Stilgenbauer
Figure 5.6	Courtesy of Rainer Schmidt Landscape Architects, graphic by Judith Stilgenbauer
Figure 5.8	Rainer Schmidt Landscape Architects
Figure 5.9	© April Philips Design Works
Figure 5.10	© April Philips Design Works
Figure 5.11	Courtesy of urban edge studio of SW+A, photo by William T. Eubanks, FASLA
Figure 5.12	Courtesy of urban edge studio of SW+A
Figure 5.13	Courtesy of urban edge studio of SW+A, photo by William T. Eubanks, FASLA
Figure 5.14	Courtesy of urban edge studio of SW+A, photo by William T. Eubanks, FASLA
Figure 5.15	Created by Jake Voit, founder of Community Earth, for Cagwin & Dorward
Figure 5.16	Created by Jake Voit, founder of Community Earth, for Cagwin & Dorward
Figure 5.17	Created by Jake Voit, founder of Community Earth, for Cagwin & Dorward
Figure 5.18	Courtesy of Riverpark Farm, photo by Ari Nuzzo
Figure 5.19	Courtesy of Riverpark Farm, photo by Ari Nuzzo
Figure 5.20	Courtesy of Riverpark Farm, photo by Ari Nuzzo
Figure 5.21	Courtesy of Riverpark Farm, photo by Ari Nuzzo
Figure 5.22	Courtesy of Riverpark Farm
Figure 5.23	Courtesy of Riverpark Farm, photo by Ari Nuzzo
Figure 5.24	Courtesy of Riverpark Farm, photo by Ari Nuzzo
Figure 5.25	© April Philips Design Works
Figure 5.26	© April Philips Design Works
Figure 5.27	© April Philips Design Works
Figure 5.28	© April Philips Design Works
Figure 5.29	© April Philips Design Works
Figure 5.30	© April Philips Design Works
Figure 5.31	© April Philips Design Works
Figure 5.32	© April Philips Design Works
Figure 5.33	Photo by Hillary Geremia
Figure 5.34	Photo by Pamela Broom
Figure 5.35	Photo by Tony Arrasmith
Figure 5.36	Courtesy of CMG Landscape Architecture
Figure 5.37	Courtesy of Sacred Heart; photo by Marco Esposito/SWA
Figure 5.38	Courtesy of Sacred Heart; drawing by SWA
Figure 5.39	Courtesy of Sacred Heart; photo by Tom Fox/SWA
Figure 5.40	Courtesy of Sacred Heart; photo by Tom Fox/SWA
Figure 5.41	Daniel Homsey
Figure 5.42	© Rebar

Chapter 6

Figure 6.1	Photo courtesy of Atlanta Botanical Garden
Figure 6.2	Photo courtesy of Atlanta Botanical Garden
Figure 6.3	Courtesy of Atlanta Botanical Garden
Figure 6.4	Photo courtesy of Atlanta Botanical Garden
Figure 6.5	Photo courtesy of Atlanta Botanical Garden
Figure 6.6	Photo courtesy of Atlanta Botanical Garden
Figure 6.7	Photo courtesy of Atlanta Botanical Garden
Figure 6.8	Photo courtesy of Atlanta Botanical Garden
Figure 6.9	Photo courtesy of Atlanta Botanical Garden
Figure 6.10	© April Philips Design Works
Figure 6.11	© AECOM 2012
Figure 6.12	© AECOM 2012
Figure 6.13	© AECOM 2012
Figure 6.14	© AECOM 2012
Figure 6.15	© Herzog & de Meuron
Figure 6.16	© Herzog & de Meuron
Figure 6.17	© Herzog & de Meuron
Figure 6.18	Courtesy of Big City Farms
Figure 6.19	© Douglas Gayeton at Rumplefarm, LLC
Figure 6.20	© City Slicker Farms
Figure 6.21	Courtesy of Terra Fluxus—Landscape+Urbanism, image by Jason King
Figure 6.22	Richard R. Kay
Figure 6.23	Richard R. Kay
Figure 6.24	Richard R. Kay
Figure 6.25	Richard R. Kay
Figure 6.26	Photo by Andrew Storey ©
Figure 6.27	Photo by Andrew Storey ©
Figure 6.28	Courtesy of CAST architecture
Figure 6.29	Courtesy of CAST architecture
Figure 6.30	Courtesy of Harrison Design Landscape Architecture
Figure 6.31	Courtesy www.permacultureprinciples.com; based on original concept by David Holmgren, www.holmgen.com.au
Figure 6.32	Courtesy of photographer, Lindsey Goldberg
Figure 6.33	Courtesy of photographer, Jessica Christian
Figure 6.34	Courtesy of photographer, Jessica Christian
Figure 6.35	Courtesy of Gotham Greens, by Ari Burling
Figure 6.36	Courtesy of Gotham Greens, by Ari Burling
Figure 6.37	Courtesy of Gotham Greens, by Ari Burling
Figure 6.38	Courtesy of Gotham Greens, by Ari Burling

Index

#
2001 Market Street, 108–112, *34, 109, 110*
    design team, 112

A
Access, 169
Advocacy, 252
Aeroponics, 162
Aesthetics, 116–117
Agriculture:
    intensive, 161
    traditional, 160
    urban
        definition, 45,48
        landscape(s). *See* food landscape(s)
Agro-ecology, 50
Agroforestry, 163
Agtivist(s), 49
Air quality, 77
Alameda, California, 209–212
Alemany Farms, 250–251, *250–251*
    design team, 251
Animal husbandry, 165, 173
Aquaculture, 162
Aquaponics, 162
    design considerations, 172
Artisan agriculture, 67
Atlanta, Georgia, 227–231
Atlanta Botanical Garden, 227–231, *227–231*
    design team, 231
Atherton, California, 218–221

B
Backyard farming. *See* edible estate(s)
Baltimore, Maryland, 38–40, 256
Banyan Street Manor, 177, *177*
    design team, 177
Bar Agricole, 10–12
    design team, 12
Beacon Food Forest, 255, *255*
    design team, 255
Beekeeping, 173
Bidding, 154–155
Big City Farms, 38–40, 67, 249, *38–40, 242*
Biodiversity, 35, 76
Biodynamic farming, 178
Biointensive farming, 166
Bioregionalism, 50
Bioremediation, 170

Blog(s), 246
Budget(s), 150
BUGA. *See* Die Plantage
Building(s), 71
    skins, 176
Business:
    model, 121–122
    plan, 103, 121–122, 149

C
Café Reconcile, 213–215, *213–215*
Calorie farming, 178
Carbon sequestration, 77
Change agent, 4–7
Charette(s), 100
Charleston, South Carolina, 75, 191–193, *75, 191–193*
Chemicals to avoid, 18
Chicago, Illinois, 117–120
Chicken(s), 173, *174*
China, 80–84
Circulation, 66, 169
City homesteading, 164
City Slicker Farms, 13–16, 215–216, 249, 260
Climate:
    control, 77
    seasonal impacts, 171
Codes, 146
Collaborative conversation(s), 147, *148*
Communications network(s), 112
Community:
    engagement, 213
    garden(s), 73, 126
    supported agriculture, 166
    table, 187
    workshop(s), 99
Companion planting(s), 178
Company food landscape(s), 131, 217
Composting, 170–171
Connectivity. *See* circulation
Construction:
    documentation, 154–155
    methods, 160
Container farming, 165
Cowgirl Creamery, 67
Cuba, 19–22, *19–22*
Cynefin Framework, 36

D
Demonstration garden(s), 73, 127
Design-build construction, 155

Design process sphere(s), 84–85, 90–94, *85, 91*
  lifecycle operations sphere, 188–189, *188*
  outreach sphere, 232–234, *232*
  planning sphere, 94–96, *95*
  synthesis sphere, 112–117, *113*
  systems integration sphere, 139
  vision sphere, 96–103, 108–109, 112, *101*
Design:
  synthesis, 115
  typology, 122–132, *125, 130*
Detroit, Michigan, 1–4
Die Plantage, 181–185, *181–185*
  design team, 185
Distribution, 68, 205–206, 241

E

Ecoliteracy, 32–33, 56–57
  center for, 57, 62
  practices, 57
Ecosystem:
  planning, 66
  services, 33–35
  strategies, 76
Eco village(s), 50–53
Ecuador, 28
Edible:
  estate(s), 128, 165
  hotel landscape(s), 127
  landscape(s). *See* food landscape(s)
  urbanism, 49
Eight Steps to Successful Change, 37
Erosion, 76
Estimating harvest, 165
EXPO 2015, 239–241, *239–240*
  Advisory board, 241

F

Farm to table, 9, 10, 127
Feedback loop(s), 147
Finance, 241
  plan, 194, 221–223
Fish, 173
Food:
  as a platform, 65
  desert(s), 12–13, 260
  justice, 12, 77, *16*
    projects, 16
  landscape(s), 49, 69–71, 76, 77, 164
  pantry garden(s), 127
  security, 12, 26, 260
  shed(s), 164, *164*
  system(s), 46–48, 59–61, *46, 60, 61*
    regional, 68
    sustainable, 69–71, *71*
Forage farming, 166
French intensive, 178
Funding, 108, 233–234
  sources, 222

G

Garden city, 8, 50, *8*
Gary Comer Youth Center, 117–120, *71, 117–120, 162*
  design team, 120
Germany, 181–185
Glide Church, 259, *259*
Gotham Greens, 261–263, *261–263*
  design team, 263
Grant(s), 234
Graze the Roof. *See* Glide Church
Greenhouse(s), 161, 206, *207*
Greenway(s), 73, 76, *74*
Greyslake, Illinois, 41–43

H

Habitat:
  corridor network(s), 76
  creation, 76
  loss, 35
Habitat for Humanity, 51–53
Hawaii, 177
Healdsburg, California, 133–137
Health, human, 77
Heat island effect, 77
Honolulu, Hawaii, *See* Hawaii
Howard, Ebenezer, 8
Human scale agriculture, 67
Hydroponics, 162
  design considerations, 172

I

Incredible Edible House, 156–159, *156–159*
  design team, 159
Infrastructure:
  green, 74, 131, *74*
  integrated systems, 66
  transportation, 73
Installation, 152, *153*
Integrated decision making, 114
Intensive planting methods, 178
Intensive farming, 178
Interbay P-patch, 253–254, *253–254*
  design team, 253
Interconnectivity, 58
Interdependence, 58
Invasive species, 35
Irrigation, 76
  technologies, 173
Italy, 239–241

K

Kotter, John P., 37

L

Labor, 208–209
Lafayette Greens, 1–4, *1–4, 72*
  design team, 4
Land farming, 161
Learning garden(s), 127
Lifecycle(s), 78, 186, *190*

operations, 186–189
performance, 149–150
strategies, 78–79
Liu, John, 26–27
Living:
  machines, 173
  wall(s), 175
Location maximization, 67
Loess Plateau, The, 26–27

M

Maintenance, 186–187
  criteria, 189
  framework, 208
  logistics, 171
  manual(s), 197
  mapping, 196–197
  plan(s), 194–196
Management, 186–187
  plan(s), 194
Marin County, California, 104–107
Market Farmers, 216
Marketing, 242–245
  plan(s), 194
Master gardener program(s), 217
Material selection
Meadows, Donella, 56
Median(s). *See* streetscapes
Medical University of South Carolina, 192–193, *192–193*
  design team, 193
Medlock Ames, 133–137, *133–137, 244*
  design team, 137
Mentoring, 213–217
Micro farming, 164
Milan, Italy. *See* Italy
Miller Creek Middle School, 102–107, 246, *73, 104–107*
  design team, 107
Mini farming. *See* micro farming
Mission statement(s), 101–103
Monitoring, 245–249
Multifamily landscape(s), 129
Munich, Germany. *See* Germany

N

Netherlands, The, 87–89
New Orleans, Louisiana, 29–31, 144–146, 213–215
New York, New York, 167–168, 202–205, 261–263

O

Oakland, California, 13–16, 260
Open space(s), 72
  design, 66
Operations, 171
Opportunities and constraints map(s), 114
Organic landscape(s), 172
Orr, David W., 32
Our School at Blair Grocery, 144–146, *144–146*
Outreach, 242–245

P

Park(s), 72, 131, *72*
Participant(s). *See* Stakeholder(s)
Partnership(s), 238–239
Peak:
  oil, 19–23
  resources, 19
  soil, 25–27
  water, 27–29
Permaculture, 51, 68–69, 255, *255*
  principles, 68–69
Permit process(es), 155
Physical planning, 65–66
Placemaking, 64–65
Planned neighborhood food landscape(s), 129
Plaza(s), 72, 131
Policy, 62–64, 252, 256–258
Portland, Oregon, 248–249, *249*
P-Patch:
  Beacon Food Forest. *See* Beacon Food Forest Interbay. *See* Interbay P-Patch Program, 66, 253–255
Prairie Crossing, 41–43, 216–217, *41–43*
  design team, 43
Program analysis, 114

R

Rainwater harvesting, 172
Recycled material(s), 176
Recycling, 77
Renewable energy, 78
Research, 245–249
  gardens, 126
Resort landscape(s). *See* edible hotel landscape(s)
Resource(s), 169–173, *124*
  collapse, 25
  cultural, 141, 142
  economic, 141, 142
  environmental, 142
  sustainable, *124*
Restaurant landscape(s). *See* farm to table
Return on investment, 151–152
River Falls Eco Village, 51–53
  design team, 53
River Falls, Wisconsin, 51–53
Riverpark Farm, 202–205, *202–204*
  design team, 205
Roof planter(s), modular, 176
Rooftop farming, 161, *163*

S

Sacred Heart Preparatory, 218–221, *218–220*
  design team, 221
San Francisco, California, 10–12, 109–112, 250–251, 259
Scale:
  aggregation, 38, 67
  development, 67
Scent of Orange, 80–84, *80–83*
  design team, 84
School garden(s), 73, 127, 217, *73*
Science Barge, The, 167–168, *167–168*
  design team, 168

Seattle, Washington, 66, 100, 248–249, 253–255
Seed to table. *See* farm to table
Self-watering planter(s), 176
Seniors, 217
Sidewalk(s), 73–75, *74*, *75*
Site:
    analysis, 114
    selection, 108
Slow food, 5–6, 9
Slow Food Nation Victory Garden, 224–225, *224–225*
    design team: 225
Soil:
    assessment, 199–200, *199–200*
    health, 25–27, 76, 170
    maintenance, 199, *199*
    management, 198–201, *199–201*
    mixes, 176
    nutrients, 170
Solar orientation, 169
Solar technology, 176
Square foot method, 178
Stakeholder(s), 123, 189, *123*
Steingraber, Sandra, 16–17
Stormwater:
    management, 76
    quality, 76
Street(s). *See* streetscape(s)
Streetscape(s), 73–74, 131, *74*
St. Regis Hotel, *128*, *129*
Sustainability plan(s), 256
Sustainable framework plan(s), 116
Sustainable Sites Initiative, 33, 65
System(s)
    analysis, 113
    connections, 140, 143
    integration, 138–144
    sphere. *See* Design process sphere (s)
    thinking, 54–57, 64, 74, *55*, *56*

T

Tool(s), 206–208, *206*
Training, 215–217
Transition:
    process(es), 23–24
    timeline(s), 23
    town(s), 23

U

Upcycling, 77
Urban agriculture landscape(s). *See* Food landscape(s)

Urban:
    design, 65–66
        strategies, 69–74
    farm(s), 126
Urbanism:
    Agricultural, 50
    Edible, 49
    Landscape, 50
Urban Food Jungle, 235–237, *235–237*
    design team, 237
Urbanite, 176
Urban Agriculture Research and Practice, The

V

Verge Sidewalk Garden, 75, *75*
Vertical:
    farming, 163, 175
    planter(s), 175, *175–176*
VF Outdoors Corporate Campus, 209–212, *153*, *209–212*
    design team, 212
Victory garden(s), 131, 224–225
Viertel, Josh, 5
Viet Village, 29–31
    design team, 31
Villa Augustus, 87–89, *87–89*

W

Waste, 77
    management, 170
Water:
    availability, 169
    conservation, 76
    reclamation, 172
Waters, Alice, 9
Wellness garden(s), 131, 132
West Oakland Park and Urban Farm. *See* also City Slicker Farms, 14–16
    design team, 16
Woolly Pocket(s), *175*
World Café, 98–99, *99*

Y

Yardshare(s), 128

Z

Zigas, Eli, 256–257
Zoning, 252, 256